COLOURS
INSIDE OUT

BEST WISHES.

WHIPPLE

COLOURS INSIDE OUT

HOW COLOUR, TEXTURE AND STYLE EQUALS PERSONALITY

Edita Whipple

Edita International Publishing
Vancouver, Canada

ISBN 0-9685148-0-4

CANADIAN CATALOGUING IN PUBLICATION DATA

Whipple, Edita, date.
Colours inside out

1. Color—Psychological aspects. 2. Fashion—Psychological aspects. 3. Personality. I. Title.
BF789.C7W55 1999 155.9'1145 C99-910380-6

Printed in Canada by Friesens Books

Cover Design: Sage Design
Text Design: Sage Design
Photography: Evan Warneboldt

Edita International Publishing
54-5531 Cornwall Drive
Vancouver, B.C., V7C 5N7

To Francis with Love, always

Contents

Introduction

FROM THE DAWN OF CULTURE TO THE PRESENT, PEOPLE HAVE been affected by colour in nature, and have developed strong associations with specific hues.

Yellow, the colour of the sun, and dark blue, the colour of night, made a strong impact on our ancestors. Early civilizations worshiped the sun. Yellow and gold meant abundance, growth and wealth. White signified purity; blue and black suggested authority; and red meant strength and power.

The 1920s were an elegant era of soft beige and creamy off-white—a decade of satin and pearls whose fashion icon was the famous Parisienne designer Coco Chanel. During the 1930s people sought refuge in rich, painterly earth tones. Khaki represented the World War II influence in the 1940s. The 1950s brought innocent pastels forward. Then the 1960s exploded with psychedelic hot pinks and Day-Glo colours. In the 1970s, rusts and browns were worn by idealitsts fighting conformity.

Burgundy red, teal and steel grey became the rage in the 1980s, which gave way to black, a non-colour conveying power and action. While we are heading to the end of the millennium, bright citrus colours dominate television and the Internet, vying for the attention of baby boomers. From antiquity to the present, each era was associated with unique colours. These were expressed first in clothing, then in home decor.

Colour influences our lives whether we know it or not. Theories of colour psychology have been developed over the years by scientists and others. Though no theory is universally accepted, psychologists agree that colours affect mood and influence emotion.

My own involvement with colour began in fashion school where I studied costume design. In theatre, costume Colour, Texture and Style are selected to convey the personality of the character portrayed, as well as to compliment the set. When I opened my own fashion school it seemed only natural to help my students develop a wardrobe and home environment which suited their personalities and lifestyles—no different than costume and set design for the stage.

Before I could do that, I had to find out more about them; their occupation, their leisure time activities, their home surroundings, their colour preferences and many other questions about their life style. We then had fun developing the appropriate Colours, Textures and Styles, putting it all together—just right for each person.

I spent the next thirty years doing just that: helping people select colours they liked and best fit every aspect of their lives. These experiences confirmed the fact that certain personalities are attracted to certain colours. Therefore, it was easy to come to the conclusion that colours are associated with specific personality traits. Over the years, my theory became so well established that peoples' colour preferences could be used to indicate specific personality traits. I also believe that changes in our lifestyle, environment or within ourself will result in change to our colour preferences.

I would like to introduce you to the concept of communicating through colour and at the same time show you how to achieve a look based on Colour, Texture and Style uniquely suited to your personality and lifestyle in complete harmony with your surroundings.

I want this book to help you get a deeper understanding of yourself, your family, friends and associates. More importantly, this book is intended to encourge you to develop an "I-care" attitude for yourself, your surroundings and life itself. I also want to show you the value of doing those little things which will enhance all areas of your life. It is my sincere wish that my message brings more harmony, balance and understanding into your world, and that you enjoy *Colours Inside Out.*

Edita Whipple

Basic Colour Personalities

The following chapters contain information gathered from my students in seminar settings and from one-on-one consultations which took place in their homes and offices over more than thirty years. This experience provided the background for my theory about colours being associated with personalities and is the foundation of this book.

If colours in your environment are not in harmony with your personality, you will feel uncomfortable. This might be the reason that you don't feel comfortable in a hotel room decorated in colours not suited to your personality. Knowing why you feel relaxed in one situation and aggravated in another will help you to correct or succumb to your environment.

Most likely, you will see yourself in one of the colour categories most of the time, and in others occasionally. You will discover the reason why you like an orderly house, or the colour **BLUE**, or a relaxed environment and what your colour preferences communicate to those around you. This is especially true when two different personalities live under one roof. I have observed that for a person to wear or surround themself with Colours, Textures and Styles not in harmony with their personality can be a strain and bring about an expenditure of physical energy.

Because my background has been primarily in fashion design, wardrobe planning and colour analysis, my main focus is on what Colour, Texture and Style of clothing communicate. I know it to be true, that when your clothes and your environment are in harmony with your personality, you have created a comfortable, restful setting for yourself, your family and friends.

Colour/Personality Scale

In order to illustrate the relationship between the Basic Colours which portray different personalities I use the following Colour/Personality Scale.

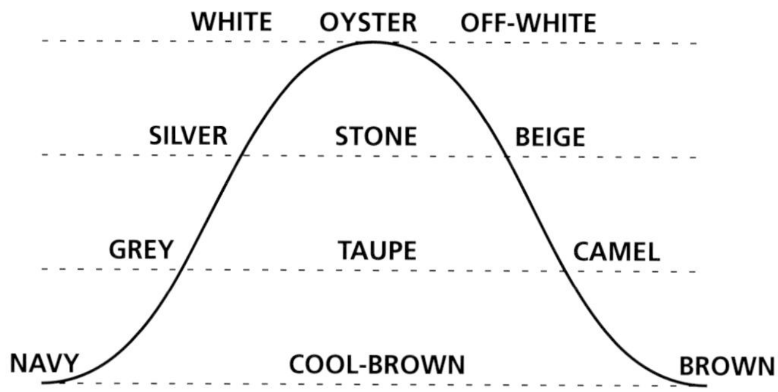

The majority of people fall between the two extreme colour personalities of **NAVY** (cool) and **BROWN** (warm). The darkest shades on the baseline are the Parent Colours which correspond to the most pronounced personality characteristic. In the text throughout the book I will refer to them as Parent Colours. Moving upwards on the vertical axis of the Colour/Personality Scale, the Parent Colours become lighter in shade, referred to as Pale Basics and the related personality characteristics also become less pronounced.

To illustrate the characteristics of the Basic Colour personalities, I chose typical examples. The following photographs were not staged but depict real people in their own environment.

I had anticipated that it would be necessary to assemble vignettes of furniture and accessories from various specialty stores to illustrate the home environments of the different personalities I wanted to portray. These had been revealed to me in interviews with students over many years. Instead, when I was looking for front doors typical of the different personalities I not only found appropriate front doors but I also found wonderful people inside, most of whom I did not previously know. These gracious people invited us in and permitted us to photograph their homes. Nothing was staged. They set their own tables and arranged their own flowers for these pictures. All the homes had character, reflected the personality traits of the owners, and had their signatures.

It was gratifying that this experience again confirmed my theory regarding colour personalties. That is what this book is all about.

The first example is of the **NAVY** home of our friends, Joan and Bill who so kindly offered me their home for "show and tell".

The Conservative Navy Personality

In general, **NAVY** represents traditional and lasting values including trust, honesty, security, responsibility, and reliability. The person who favours **BLUE** to other colours wants an orderly life and an environment free of disturbances. The **NAVY** personality strives to achieve calm through activity, has strong emotional control and a high level of concentration. These individuals are resourceful, know how to achieve financial success and are especially concerned about their economic security.

The **NAVY** personality traits are found in most people to one degree or another because of the society in which we live. Desire for order can be traced back to traditional values to which most of us can relate. This chapter portrays the typical, traditional personality in their surroundings and attire.

My observations made over many years indicate that the persons who favour **BLUE** generally prefer a traditional home—starting with an aesthetically pleasing front door. I was always assured of finding beautiful rooms inside. The traditional house, townhouse or apartment has a well-kept front yard. The need for order is high on the list for persons who like **BLUE** and if they are not able or willing to create an orderly environment, they will usually find an assistant (or spouse) to do it for them. "A place for everything and everything in its place" is the motto of the **NAVY** person. Accumulated chores and disorder usually results in irritability and loss of energy. Visitors are wise to phone before dropping in or, better still, wait to be invited by the host/hostess at a time when they have everything in order. The true **NAVY** personality is not tolerant of dogs or cats that shed hair, or partners that create disorder.

Open the Door to the Navy Household

5091

The Navy Living Room

Welcome to a beautiful living room. Traditional furniture is the mark of a **NAVY** personality. Time-honored pieces not only grace the room but are considered to be a good investment. Beautiful antiques, paintings and etchings are displayed. Oriental carpets on parquet floors, or thick wall-to-wall broadloom in pastel colours adds its own touch of luxury to the room. The furniture is covered in shimmering damask, soft velvet or sateen. Crystal chandeliers and lamps topped with silk shades illuminate the room. Some **NAVY** individuals may prefer an English country atmosphere complete with a brick fireplace, chintz-covered Chesterfields and batwing arm chairs. Polished brass and copper accessories enhance their traditional living room. Not all **NAVY** personalities decorate with antiques only, but there is always an heirloom piece or two, a chest of drawers, an armoire to hold the sound system or a beautiful beveled mirror to set the stage for the **NAVY** personality who lives here.

The Navy Dining Room

The formal dining table is set with lovely linen, china, cut crystal and ornate silver flatware. Others might use lace-edged placemats and napkins on a highly polished table graced by silver chargers to accent the fine, plain-white china and uncut crystal wine goblets. The **NAVY** personality who prefers the more traditional country style most likely will use a crystal of substance, such as Waterford. The heirloom china is hand-painted and they might use pewter candle holders and napkin rings, instead of polished silver. This personality will use a linen table runner and matching placemats and napkins on an oak table. Fresh flowers and candles enhance these beautifully appointed rooms even further.

Flower Arrangements

Flower arrangements make another statement of the different personalities. As you see in the pictures, the **NAVY** person prefers roses, usually red or pink, displayed in crystal, china or silver vases (never in containers not intended for flowers). Several of my **NAVY** students and clients prefer carnations for their long-lasting quality and neatness. They don't shed petals. *NO!* I did not make this up!

Kitchen

This work area is orderly, void of clutter on the counters, and the dishes are always done before retiring. Some students told me that they even vacuum their sink. That's better than neat! Containers that hold sugar, flour, coffee or spoons are clearly labeled and tea towels are ironed.

Food

Meals emerging from this kitchen remind one of the traditional dinners eaten back home. Roast beef, Yorkshire pudding and apple pie for dessert (or its equivalent according to the person's origin). The new generation may follow the advice of nutrition gurus but comfort food is still a meal that evokes childhood memories and is most satisfying. The longing to share meals with family is especially strong at Thanksgiving, Christmas and other special occasions.

Occupation

Because the **NAVY** personality likes a structured life, they have a tendency to choose careers in the field of science, engineering, law, banking or accounting. As we know, **BLUE** is the colour of the establishment.

Activities

The **NAVY** personality is often drawn to classical music, and loves opera, ballet and theater. This person is also the one to raise funds to help advance the cultural life in their community and they contribute to charities. Dinner away from home is usually shared with one special person or a few friends of long standing, and takes place in beautiful restaurants that are known for excellent service. A private club is often chosen for entertaining larger groups of friends or business associates. The **NAVY** personalities prefer to spend holidays in cosmopolitan cities or to spend their leisure time in climates opposite to that of their home town.

Sport

Highly disciplined sports such as jogging, aerobics, downhill skiing, racquetball, tennis, cricket, riding, sailing and golf are favourites of the **NAVY** personality. The outward signs of personal achievement are evident by belonging to a private club, owning a boat, a ski cabin or an antique car. The feeling of satisfaction derived from achievements provides them with a sense of peace which is of great importance to the action-oriented **NAVY** personality in their quest for order and tranquility.

NAVY Summary

personality = conservative
environment = traditional
activities = action-oriented

Colour/Personality Scale

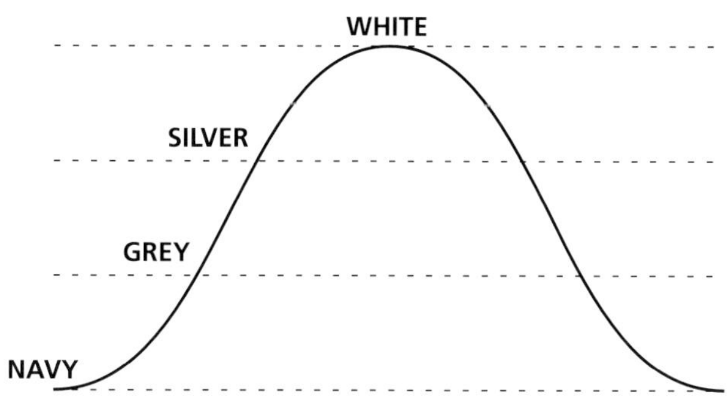

On the baseline, the Parent Colour **NAVY** is the deepest colour and portrays the most pronounced characteristics of the conservative personality. **GREY** reflects the characteristics of **NAVY** but has a more flexible attitude. **SILVER** is more relaxed and open to different experiences but never forgets the Parent Colour's personality traits. **WHITE** is the lightest shade and communicates crispness and preciseness.

Not all people who are conservative, like traditional furniture, and set their table with silver and china have to fall into the norm in the way they dress. The interior of the **NAVY** home in the illustration belongs to a couple I have known for several years. Joan and Bill were clients of mine and subsequently became close friends. Joan offered me the use of her home for these pictures—nothing was staged. She set the table and arranged the flowers the way she always does for their friends and family. Joan and Bill both are **NAVY** by background. Their home is traditional, the living room and dining room are graciously and beautifully appointed. Joan and Bill belong to a private club and both like the colour **BLUE.** They spend a lot of time in their cottage on a nearby island and have decorated their retreat in **BLUE** and **WHITE** throughout. Both our friends enjoy a game of golf and Joan also plays tennis. So far everything falls into the category of **NAVY** personalities except that they both like a casual, unconstructed lifestyle, love being with their friends and always have time for them.

Here are my suggestions for dressing in harmony with their personalities and lifestyle: both have **NAVY** and **SILVER GREY** as their Basic Colours, and Joan wears a lot of raspberry-red and bright blue. Bill wears bright blue for his sport shirts and turtlenecks, topped by a sweater for texture over grey flannels or twill pants. Since retirement he rarely wears a tie. Joan's style of dress is unconstructed. She wears beautiful linen blouses, loose pants, and full skirts which are mostly anchored by a traditional style of jacket in bright colours. Bill and Joan's looks are ones of ease, executed in their favorite colour scheme—**BLUE**.

You may notice that I try to soften the **NAVY** person's image and their structured outlook on life in general, by portraying a more open attitude for family, friends and business associates through their choice of clothing. I also invite the **BROWN** personality to add some structure to their image.

The Navy Clothing

When you are looking at clothing to suit your personality rather than colours which match your hair and/or skin tones, you will need a little mind shift. When I talk about a **NAVY** person or a **BROWN** person, I mean you, who fits into one of the lifestyles and environments described in this book and who also displays a certain style of dress using Colours, Textures and Styles which communicate your personality.

Most of my closest friends relate to the colour **BLUE** in all shades and exhibit the corresponding personality traits to varying degrees. All enjoy an orderly lifestyle and have aesthetically pleasing surroundings (which has nothing to do with finances). They are action-oriented people. A game of golf or tennis or a hike in the mountains give them a sense of achievement. Then they are at ease and feel they can afford to relax.

During my years of teaching and consulting I have experienced more **NAVY** personalities than any other colour. However, the reason for this could be the **NAVY** person's willingness to seek professional advice.

Knowing something about your surroundings and your way of life is the first step to determine the appropriate Colour, Texture and Style of clothing in harmony with your personality. The choice of the colour **BLUE** is a good starting point to portray the personality of the traditional and/or conservative person. **NAVY** represents the extreme personality traits. However, for the less conservative person, a lighter colour of **BLUE** is appropriate or moving up on the colour personality scale to **GREY** or **SILVER**.

The basic style of dress appropriate for the **NAVY** personality conveys control and is always in good taste. Their preference is usually toward clothes which are structured and tailored to perfection. This style of dressing communicates that the person is efficient, self assured, trustworthy, reliable, orderly, business-oriented, capable and most probably does not squeeze the tube of toothpaste in the middle. A person dressed in strict **NAVY** style does not invite a shoulder to cry on but is a great friend for giving advice on how to manage your budget or clean your closet. This person probably works with a time system and you'd better make an appointment for coffee or lunch.

The fabrics worn by **NAVY** personalities are of good quality and must have wrinkle-free properties: flat-finished fabrics such as twill, gabardine, worsted wool, silk crepe and crisp cottons (blended or treated to make them wrinkle-free). Regardless of the fabric, the **NAVY** person always looks well groomed.

What I would like you to see is that by choosing your Colours, Texture, and Style, you can convey a message for different situations and surroundings. The Costume of the true **NAVY** person is shown in the picture. If you are a **NAVY** person, you will relate to it; actually a lot of people do—**NAVY** spells success. This image can also create distance when dealing with people and you will have to spend some extra energy if you want to convey a message of "being down to earth." But, if you want to convey efficiency and the "I get things done" look—this is it.

Colour + Texture + Style = Personality

NAVY + Worsted + Close to the Body = Conservative

e

Gems from my Collection

This gem about colour changes to portray a different image happened when Peter attended one of my seminars for men. He was so delighted with his new image, that he went home and piled all the clothes which were not part of his new plan in the middle of the floor, never to be worn again. He had favoured all unconstructed styles in **BROWNS**, **BEIGES** and **EARTHTONES.** For his role as mayor, we decided on **NAVY** (to **GREY** and **SILVER**) as the appropriate colour, but kept the texture of the **BROWN** personality by using tweed fabrics and sweaters.

Peter's style is traditional but he gives it his own signature with the use of red accessories, e.g., sweaters, flannel shirts, ties and belts. Pale blue or white striped shirts give polish to his blue tweed suits, tweed jackets, and flannel slacks. He carries his colour scheme through to his leisure activities. He hikes and skis in navy and red attire—no longer beige or brown. No matter what he wears, his look always conveys his relaxed signature in dress.

His style is that of a country gentleman: tweed suits, flannel jacket and pants and a navy blazer that he wears with shirts in small checks or a red V-neck sweater. Peter dresses with ease and a natural flair suitable for his position.

Navy to Beige

Henry arrived at our meeting in proper business attire—the generally accepted navy blue blazer, grey slacks, white shirt, wine red silk tie, black shoes and socks, and a navy blue raincoat. He is the owner and manager of a travel bureau and likes the rough outdoor life, spending much of his spare time in a rustic cabin in the country. He goes fishing and skiing. His favourite piece of clothing is a handknit beige sweater. Henry likes people and is dealing with them daily in this business. It is his responsibility to create holiday packages that inspire his clients to travel. He has to convey to them, as well as to business travelers, that their arrangements are handled properly and that their journey would not be encumbered with delays or problems.

To give the impression of his sincere interest in his clients, his attire needs to look relaxed and communicate caring. Brown corduroy slacks and a sweater would be fine for that, but not business-like enough and too relaxed for his corporate clientele. Because he is the owner of the business, his attire sets the example for his company image. We decided on a less structured image that his holiday customers could relate to, and combined it with the accepted look of a businessman. The navy blue blazer and pale blue shirt identified him as a businessman to his corporate clientele and the change from grey to beige for slacks and socks, and from black to tan shoes and belt, made the appearance less intimidating to his holiday customers. The red textured tie dotted the "i."

Grey to Beige

My husband, Francis, worked in the corporate world most of his business career and wore the appropriate and expected clothes. Navy raincoat, navy and grey suits, white shirts and burgundy or blue striped ties. He chose black shoes and socks with his blazer and grey flannels, of course. He retired at a young age, and when I took the first picture of him—slim, trim and tanned, holding his catch, a big salmon, in his hands—I realized how he had changed. No longer did he fit the corporate image. I rearranged his closet and gave away his grey suits, grey flannels and burgundy sweaters. Francis now wears beige suits with pale blue shirts and beige wool slacks with his navy blazer. His accessories are red rather than burgundy, ties with patterns or texture and his shoes are mostly tan or brown. The only time I see him in navy or black shoes is when he wears his navy suits. I put a heart-shaped card in his pocket which reads: I love you and your new image.

The Unconventional Cool-Brown Personality

The characteristics of a **COOL-BROWN** personality include a strong desire to be unique and original, and a distinct dislike for control in spite of a conventional background. These people are not intrusive but rather soft-spoken. They are charming, have a primary attraction to beauty, and are given to fantasy and daydreaming. The **COOL-BROWN** personality longs for something interesting and exciting to happen in their life. Mediation, metaphysics, and religious studies are often their interest. Many of my **COOL-BROWN** clients use the softer **TAUPE** image. I have found that over the last decade, many of the basic **BROWN** personalities who fitted into the **BROWN** lifestyle (see next chaper), changed to the more sophisticated **COOL-BROWN** or **TAUPE** image.

The **TAUPE** environment has always been a happy surprise for me. I could not tell what was behind the door, only that I would be looking at a collection of interesting things. Because the **TAUPE** personalities are non-conformists and strive to be different, their environment and surroundings are also different and unusual in every aspect. Nothing is structured or done to a norm. I hope that we all can learn from this unstructured trait or at least open our mind to it. These people all have one thing in common: their environment is not conventional.

I had found interiors of my **TAUPE** clients and friends, but they happen to live in townhouses or apartments, with white entrance doors. Given the opportunity, wouldn't you choose a front door in harmony with the interior?

When I was looking for a typical **TAUPE** front door example, I met Karen. She opened the door and I was not surprised. Her whole house was filled with the most unusual things. She is a designer of children's clothing, and has done interior design and flower arranging. Her creativity spills out everywhere.

Open the Door to an Interesting Environment

2316

The Cool-Brown Living Room

Every room these individuals occupy is eclectic. An antique piece or two has a special place in a room enhanced by unusual wallhangings, paintings and sculptures—modern or antique. The walls are lined with books of all kinds and often one will find a piano, or other musical instrument in a corner. There are comfortable seating arrangements, and lots of pillows are tossed on couches, on Oriental rugs or the floor.

You will see conversation pieces of all kinds which these creative people found on their trips to antique dealers, flea markets or around the world. I have come across an antique guilded chair assigned for a collectors doll, a tribal wedding dress on the wall and a three foot replica of a pink volkswagen on the coffee table. Unfortunately we could not photograph all of Karen's extraordinary living room, but the corner shown in the picture is what one can expect from a **TAUPE** personality's environment.

Travels!

The Cool-Brown Dining Room

The dining room of the **COOL-BROWN** or **TAUPE** personality is furnished with a variety of different pieces which convey the owner's special style. Their environments, just like their wardrobes, are a blend of personality traits, and the dining room is no exception.

The room conveys a feeling of airiness and light. The dinner table is often topped with a piece of clear glass or made of a beautiful light wood—natural or painted. Table settings tend to be formal.The **NAVY** personality would use an antique lace tablecloth, placing it smooth, straight and orderly across the length of the table. The **TAUPE** person may use the same cloth but will drape and shape it in soft folds in a carefree way. Perhaps an Indian shawl acquired on a trip could be the background for an array of conversation pieces ranging from glass birds nestled between pink roses in silver bowls, to candlesticks with dangling stars and moons that match the pattern of the napkins. You may also find a beautiful book on a stand, its pages open to a picture which is a reflection of the meal's theme. Sometimes even a stuffed animal appears on the scene as in the photo of Karen's dining table.

China place settings may be plain but are usually in unique shapes. Plates can be a collection of pieces found here and there while traveling, or they may be inherited treasures. Candlesticks in a variety of patterns and shapes add additional charm to the table and enhance this aesthetically pleasing setting.

Flower arrangements

The containers for the display of flowers are as important as the flowers themselves. Consideration is given to colour and shape of the arrangement to make it an artistic accessory to the whole environment—again, a conversation piece. Fragrant potpourri in crystal bowls is placed in every room to add to the decor.

Kitchen

Utensils are part of the decor. Baskets of all sizes and shapes adorn the walls and hang from the ceiling. Jars of herbs and spices are displayed on shelves, giving the kitchen a rather Bohemian look.

Food

The **TAUPE** personality prepares meals with flare and uses recipes collected in their travels and from their interesting friends. Health food, vegetarian, or exotic dishes from different cultures are often prefered.

Occupation

Because of their sensitivity, the persons that are drawn to the **TAUPE** colour are usually interested in writing and painting. Sculptors and musicians feel comfortable in these "cut colours" as artists call them. University professors, especially those associated with liberal arts, as well as religious leaders, choose to wear **TAUPE** which is less threatening than **NAVY**, in order to communicate their message without creating a barrier. Psychologists and human development counselors are other professionals suited to wear **TAUPE** colours.

Activities

Being with friends who share their interest in food, dance, music, theatre, art and the literary world attracts the **TAUPE** personality. Talks and lively conversations until the early hours are common events in their lives.

Sport

Because of their artistic nature, **TAUPE** personalities prefer classes in modern dance, ballet, or Tai-Chi rather than the action oriented sports of the **NAVY** personality.

COOL-BROWN Summary

personality = unconventional
environment = eclectic
activities = meditative

Colour/Personality Scale

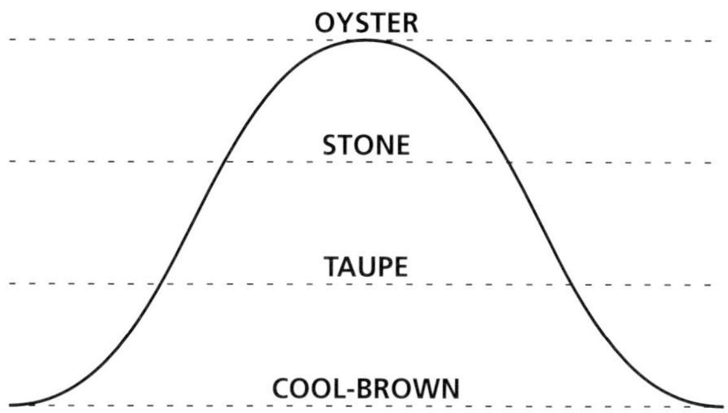

The Parent Colour is on the baseline.

- **COOL-BROWN** is the darkest colour and represents the strong desire to be unique and to break away from control.
- **TAUPE** accepts the Parent Colour traits but to a lesser degree. It is easier to communicate with a **TAUPE** image.
- **STONE** is relaxed and open to all new experiences.
- **OYSTER** reflects an even more relaxed nature.

The Cool-Brown Clothing

When planning the wardrobe for the **COOL-BROWN/TAUPE** personality, it is an advantage if they have collected some conversation pieces (clothing, accessories, jewelry), which is usually the case. These unique items contribute to a style which rarely can be duplicated and which makes a statement all of its own. **COOL-BROWN**, **TAUPE** or **STONE** colours provide a good background for displaying a collection of accessories and unusual jewelry.

Most of my **COOL-BROWN** clients are in the business world and accordingly have a somewhat **NAVY** background or lifestyle. I advise my clients to wear one garment, i.e., a tailored jacket (no collar or lapels) with their signature silk or lace blouse and flowing skirts to relate to the conventional business community. The jacket will anchor their unconstructed style of dress. Most importantly though is their colour preference. I have not met one of my **COOL-BROWN** personalities who would ever wear **BRIGHT RED**. Everyone of them favours **SOFT TAUPE**, **STONE**, **SAGE GREEN**, **SOFT YELLOW**, or **GREYED PINK** for their clothing and/or surroundings. The fabrics of these costumes are most often wool or silk crepe, to create the soft and flowing effect to which the **TAUPE** personality is drawn.

It is the colour of your clothes, from **COOL-BROWN** to **TAUPE** and **STONE** as well as your collection of conversation pieces which will more likely separate you from the other personalities.

Colour + Texture + Style = Personality

COOL-BROWN + Crepe + Flowing = Unconventional

Gems from my Collection

I met my new client, Margaret, at an art class. She wore navy pants and a white shirt—the efficient **NAVY** image. When I arrived at her house for a private consultation, I was not surprised to find a **NAVY** household. Everything was beautifully appointed. The living room had dark green walls offset by white woodwork, a brick fireplace, Oriental rugs and brass lamps. The dining room displayed antique oak furniture. When I saw her closet, I was speechless; it felt as though somebody had turned off the light. The overall effect was **GREY**, **TAUPE**, **MOSS-GREEN**, **BURGUNDY**, **BLACK** and a lot of **NAVY**. I asked her whether she was grieving or hiding.

She had engaged my services because she knew that I used the psychology of colour for planning a wardrobe. She confessed to having been in mourning for the last four years because she had been struggling with the decision to give up her private practice.

We agreed to give her a happier image. Stored away in another closet were some **BRIGHT GREEN**, **CLEAR BLUE** and **OYSTER** shirts, sweaters, blouses, pants, scarves and beautiful, embroidered belts. We even opened her jewellery boxes to discover some interesting pieces she had collected from all over the world. We mixed and matched the **GREENS** and **TAUPES** in her closet and got quite a lift. It took a while, but after two-and-a-half hours she looked in the mirror and admired her more youthful and happy image. She wore taupe pants, a clear green silk blouse cinched at the waist with an interesting belt found on her recent trip. To that we added great silver earrings and silver sandals—simple, but great looking. She is rather gorgeous to begin with, which was hidden under a lot of **GREY**.

We chose **NAVY**, **TAUPE** and **OYSTER** as her Basic Colour scheme—**NAVY**, because of her background and business involvement, and **TAUPE**, because of her soft spoken personality. Her accessory colours range from clear bright to pale blues and greens to give her a youthful and happy apperarance. I suggested a mix of crisp and soft Texture and Style, i.e., a flowing crepe skirt and a tailored jacket or gabardine pants with a chenille sweater. Her environment suited her, but I recommended to warm it up with tapestry pillows and a display of more of their collectibles to indicate that her husband lives there, too. I am told that he loves people. I understand he is a **BROWN** personality and would probably appreciate some textured accessories in their living room.

The Casual Brown Personality

Many people are drawn to the warm colour of **BROWN** for comfort. It has a homespun feeling to it. **BROWN** in different depths of tone reminds one of the outdoors, being close to nature, where life is relaxed and informal. The **BROWN** personality's desire for a quiet life is quite pronounced. Order (in the **NAVY** sense) is of little importance to this individual. Money is not a subject for discussion, but people are. The person who likes **BROWN** is a friendly, cheerful and well-adjusted person (or at least acts that way).

Being in the company of my friends who live in a **BROWN** house is very relaxing for me. They always have time to sit and talk. I chuckle while I stumble over the rubber boots at the front door (mine are in the laundry room on a rack—**NAVY**). I treasure their friendship, and admire their casual lifestyle.

The **BROWN** personality prefers to live in the country or a rural community. They prefer to have land around them—at least enough to plant a garden. This personality prefers a house with a large, comfortable kitchen or family room with enough space for friends.They prefer the outdoors and take every opportunity to escape the city to be at their country retreat, their island cabin or to go hiking.

Deborah's front door was just what I was looking for to portray the entrance to a **BROWN** personality's house. I phoned her and told her about my project. She laughed and said that all the traits I attributed to the **BROWN** image were true for her.

When we met the next day for the first time, she wore a long denim skirt and a colourful sweater. The coffee pot was on the stove and she was in the middle of preparing ten pounds of chicken for one of her children's school function. The dog wanted to play ball. I knew that I had arrived at a **BROWN** household just as I had anticipated.

See the Inside of this Welcoming Home

1968

The Brown Living Room

The floor is usually wood or covered with rustic tiles; handwoven or braided rugs give it warmth, and the comfortable seating arrangements are usually covered in textured fabrics. A stone or brick fireplace with a crackling wood fire adds to the kind of homey ambiance these **BROWN** personalities love to create. Several of the accessories reflect the hobbies of the **BROWN** person who lives in this environment; large handmade quilts are tossed on the couch, chairs or hang on a wall. Wood-carvings, restored antique benches and chairs decorate the room, and baskets of dried flowers make this a warm, informal and friendly place for family and friends to share.

The Brown Dining Room

The style of the dining room is informal and has an inviting, warm atmosphere. The large pine or oak table is set with textured placemats and napkins, a handwoven tablecloth or even a plaid blanket. Ceramic tableware collected at fairs, or inherited painted stoneware dishes invite family and friends to a country-style meal. In this environment food is served in pottery or wooden serving bowls family-style. Stainless steel cutlery with wooden handles and glasses with ceramic bases complete this **BROWN** table setting. A basket of flowers might sit on the table to start, but would be removed to make room for the soup pot. Candles in ceramic, wooden or brass holders are lit for every meal. This same setting could also be al fresco: a picnic table, benches, a hearty meal prepared on the barbeque. Most people are familiar with this setting and enjoy being part of this welcoming atmosphere.

Flower Arrangements

The flower arrangements for the **BROWN** personalities are usually as unconstructed as their lifestyle. Great bunches of wildflowers in pitchers, branches of blossoms in milk cans, a bucket full of daisies or a beanpot stuffed with evergreens and apples are displayed on tables or shelves. Even vegetables may become part of the arrangements.

Kitchen

The kitchen has lots of things everywhere: baskets, crocks, open shelves displaying preserves and cookbooks, a soup pot on the stove, pots and pans hanging from a ceiling rack. The refrigerator door is decorated with children's pictures and art; dishes accumulate in the sink; and potted herbs blossom on the windowsill.

Food

Meals prepared in this kitchen are most likely to be wholesome and generous in quantity. Soups, stews and casseroles throughout the year and a big turkey or goose for holiday gatherings are served to all who want to share in the occasion.

Occupation

BROWN is the colour personality which relates to people. The **BROWN** person chooses an occupation that offers a relaxed, unstructured life style and concentrates on people and their emotional needs. For these reasons they are drawn to professions in the humanities and social services, e.g., teachers, doctors, nurses and counselors. Those who are attracted to the outdoor life may become landscape designers or may operate a nursery or flower shop.

Activities

For the **BROWN** personality, activities are of a more tranquil nature and they enjoy a slower pace. These individuals like and support community theatre, folk music and country fairs. Many of them take up handcrafts like quilting, embroidering, knitting, woodworking or making pottery or stained glass pieces. Flower or vegetable gardening is another activity that brings these **BROWN** individuals much enjoyment.

Sports

Lots of **BROWN** personalities are "allergic" to sports, especially active sports like jogging. They prefer to go on long walks and hikes in the country. They like canoeing, kayaking or fishing from a row boat.

BROWN Summary

personality = casual
environment = rural
activities = relaxed

Colour/Personality Scale

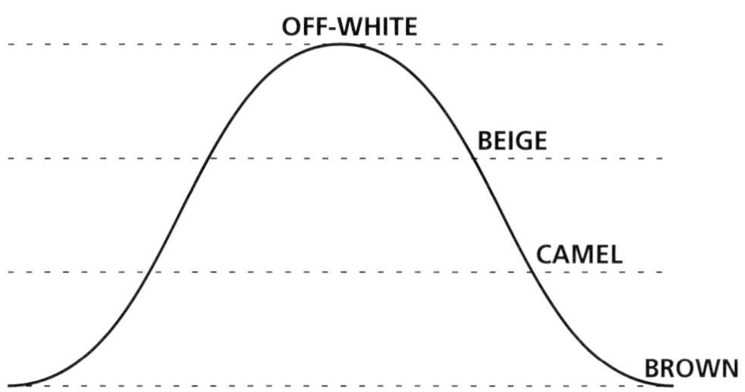

The Parent Colour and strongest personality trait is on the baseline.

- **BROWN** portrays the most relaxed, unconstructed and easy-going image.
- **CAMEL** accepts the value of the Parent Colour but makes their image convey an open attitude toward change.
- **BEIGE** is receptive to all new experiences and is a relaxed easy communicator.
- **OFF-WHITE** communicates non-involvement.

The Brown Clothing

The person who lives in the country or outlying areas of the city and spends their free time outdoors, generally takes into account the practicality of clothing. **BROWN** is a good choice, but so are all the related colours right up to beige. It just depends on how **BROWN** the personality is.

The fabrics worn by the people-oriented person are usually warm to the touch. Tweeds, wool, linen, as well as corduroy, knits and flannel invite one to touch. **BROWN** personalities do not worry about wrinkles in their garments and are at ease with that lived-in look. The **BROWN** person's relaxed and comfortable appearance makes them the choice personality to ease tension among friends and business associates. The **BROWN** wardrobe just needs a bit of structure to gain acceptance in the business world without foregoing comfort.

BROWN communicates that the person is easy to talk to, flexible and readily accessible. Their preferred lifestyle is unstructured, and, yes, they probably do squeeze the tube of toothpaste in the middle. If you have any personal challenges in your life, see the **BROWN** person for understanding support.

The basic style of clothes that I suggest for the **BROWN** personality who likes to dress in harmony with their personality—and conveys their friendly and cheerful nature—consists of unconstructed garments and fabrics which have a "soft hand." Add to these a beige tweed or houndstooth hacking jacket to anchor the look. To complement the jacket, I would choose drapeable fabric for a soft blouse rather than a shirt. Sweaters are a natural for this style of dressing and so are classic-cut pants in twill, wool or corduroy. Skirts for the **BROWN** woman are made of cotton twill, wool or denim with enough room for walking. All these garments should be of high quality. Purchase only top grain leather bags and shoes to complete the look of an outdoor-loving personality. When she is in the country, she most probably will wear jeans, T-shirts and hiking boots.

There were other **BROWN** personalities in my seminars who fit the outgoing, open-door-policy personality traits, and who preferred a relaxed lifestyle but who did wash their dishes every night (**NAVY**) and were not outdoors-oriented. These students and clients wore the same cut of clothing as described above—loose and unconstructed. Their personality was influenced by some **NAVY** traits, either from their upbringing or by their inherent nature to have things in order. Therefore they are more concerned about having their house in order before they open the door to friends, but make no excuses if it is not so—the true **NAVY** person would.

The picture of **BROWN** clothing shows this influence of the **NAVY** traits. Tweed, denim, corduroy, and a textured sweater are typically **BROWN** but the style of jacket citifies the look. The true **BROWN** personality would not wear lapels on their jacket, too structured for them, but they love the boots.

Colour + Texture + Style = Personality

BROWN + Tweed + Unconstructed = Casual

Gems from my Collection

I am sharing with you the following gem as an example that **BROWN** does not always mean that you have an unconstructed environment and lifestyle. My friend, Doris, is a dress designer and has been my business associate for many years. She has an outgoing personality, and enjoys a relaxed lifestyle. No stress! Everything in her home is orderly which indicates her **NAVY** personality traits. She does not wear or decorate with any kind of **BLUE**. The colour suited to her personality is **BROWN**. The fabrics of her clothing have a smooth texture (**NAVY**) and her style of dress is unconstructed (**BROWN**). Her Accent Colours are **TOMATO RED**, **WARM YELLOW** and **GREEN**. Her environment background colours are **OFF-WHITE** and **APRICOT** and her furniture is traditional. She uses fine china and silver on her dining room table. As you can see, not all **BROWN** personalities wear tweed and set their table with pottery. Watch for refinement of fabrics and style in wardrobe and environment which points to the **NAVY** influence. You will be able to tell the difference.

Brown to Taupe

Let's look at a change of colour dictated by a change in lifestyle. Anne lived with her husband and family in the country, along with horses, dogs and cats. Her trips to the city were infrequent. When I met her she was involved with her art and the artists in her community. I had created the look of Country Casual for her in colours of **BROWN** and **CAMEL**. The fabrics were often handwoven and we used fall colours as accents. Her home was furnished in the traditional style of the **BROWN** personality.

She and her husband furnished a new townhouse in the city with modern furniture, all in **OFF-WHITE** to provide a background for their art collection. Ann's new activities, and lifestyle required a new personal image. Her colours are now **TAUPE** and **STONE** (her warm colours became citified). The fabrics changed from heavy tweed to gabardines, and her warm Accent Colours changed to the off-colours of **GARNET**, **SAGE GREEN** and **MISTY BLUE**.

The Dramatic Black Personalities

The darkest colour and, in fact, the negation of all colours is **BLACK**, the last one of the dark Basics. **BLACK** expresses the desire to be in control of one's own decisions and actions. This personality is strong-willed, outgoing and wants to be independent and free of limitations other than those that are self-imposed. The **BLACK** personalities believe that life has a lot to offer and that they are competent to control their own destiny in any chosen field of endeavor.

BLACK denotes justice and is the colour of formality and mystique. Many people from Asian cultures wear **BLACK**, which suits their formality in behavior. In order to wear **BLACK** you have to be strong enough to overcome the effect of "no colour." You have to be self-assured and not easily intimidated.

Today young people, particularly adolescents, often choose **BLACK** as a symbol of their independence and their ability to make it on their own. While they rarely communicate the self-assured, sophisticated nature inherent to the image of **BLACK**, they are using it in an attempt to communicate exactly those qualities.

BLACK is, of course, also worn traditionally for solemn occasions—in court, at graduation and funerals, as well as being the established "uniform" for men for all formal evening occasions.

I will discuss the "little black dress" in a later chapter.

Open the Door to the Glamorous Black Environment

7650

The Black Living Room

For the **BLACK** personality, the environment is likely to be one where there is a lot of action. These persons choose to live in the center of the city or within easy reach to all kinds of activities in order to stimulate their constantly busy minds. They might have a place in the country to unwind, but more often than not, they invite their friends and the retreat becomes another place of activity.

These individuals have elegant surroundings. The furniture in their homes ranges from totally modern to antique collector's items that are their signature.

Several **BLACK** personalities I have met do not use a professional decorator, but like to create their own living space in a grand style. Oversized seating arrangements, upholstered with either silk damask, plain cotton or chenille, are the focus of the living room, no matter what size the room may be.

Accessories are also generous in size. Large pillows, lots of tall candles in either antique silver, glass or iron candleholders and large flower arrangements give the environment the feeling that every day is a special occasion.

The Black Dining Room

The dining room for the true **BLACK** personality has to be large enough to seat at least ten guests around a formal table. All my clients who fall into the **BLACK** category like to give parties. I know of a person who insists that it is perfectly fine to invite one guest per ten square feet of living space. So, if you have a thousand-square-foot living space you can entertain a hundred people. Apparently you do not have to move your furniture out. I am too **NAVY**, and therefore I am not going to try it.

As you can see, everything is generous even in the way **BLACK** personalities give a dinner party. There is nothing timid about them. Nobody is surprised to find the dinner table set with beautiful crystal and china. I have seen a table covered with volumes of damask all the way down to the floor, admired an antique highly polished table adorned with fragile organdy placemats, and a heavy beveled glass-topped table gleaming with a generous display of rare china and crystal.

The table settings of the **BLACK** personalities, no matter which Parent Colour they belong to, are the most dramatic of all.

Flower arrangements

The flower arrangements planned by a **BLACK** personality are always huge, tall and generous, or they can be the other extreme and consist of one slender single stem of an orchid, or even a tall rose in a silver bud vase. I believe these **BLACK** personalities would not know what to do with a posy of forget-me-nots.

Kitchen

The action-oriented **BLACK** personality prefers a streamlined kitchen equipped with state of the art appliances, and permits no clutter on the counters.

Food

The meals are most likely ones which are easily prepared, are planned in advance and observe the latest nutrition guidelines. The food in a **BLACK** household is always presented in an aesthetically pleasing and creative manner.

Occupation

BLACK, the colour of justice, is worn by judges and attorneys. Because **BLACK** communicates power and action, entrepreneurs prefer it in order to make a statement. Business people, on all levels in the corporate world, feel in control when wearing **BLACK**. The **BLACK** personality is a great organizer and knows how to delegate. People in sales portray a certain assertiveness dressed in **BLACK**. Fashion designers are drawn to **BLACK** for drama, as well as using it for a backdrop against which to display colours and accessories.

Activities

The **BLACK** personalities like to be involved with people. They like theatre, musical events, art exhibits and dances, as well as parties, that bring them in contact with exciting and stimulating people.

Sport

Preferred sports often are centered around a clubhouse setting. They love skiing (downhill for the thrills) and the companionship that can be found in a beautiful ski lodge après ski. They like tennis, golf and power-boating as well as flying.

BLACK Summary

personality = dramatic
environment = current style
activities = varied and action-oriented

The above summarizes the personalities of the most dramatic, outgoing and action-oriented people I have had the pleasure to work with and call my friends.

Colour/Personality Scale

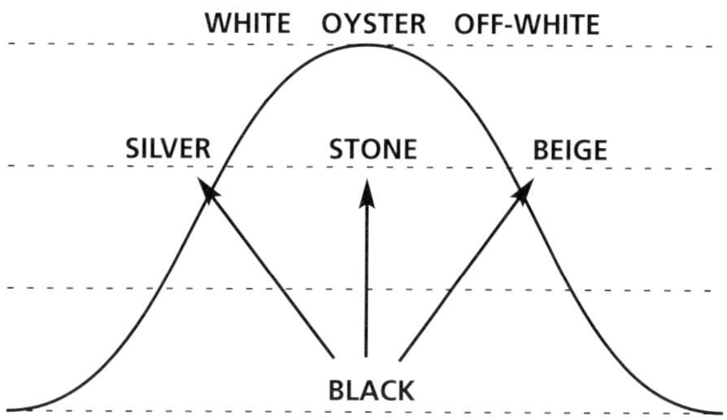

A separate Colour/Personality Scale is used for **BLACK** personalities because of their relationship to each of the three Parent personalities **NAVY**, **COOL-BROWN** and **BROWN**.

- **BLACK–SILVER–WHITE** related to **NAVY** portrays the most traditional and structured, classic and elegant image.
- **BLACK–STONE–OYSTER** related to **COOL-BROWN** is the least constructed style of the black category and conveys an image of originality.
- **BLACK–BEIGE–OFF-WHITE** related to **BROWN** stands for the self-assured outgoing personality with style.

The Three Black Colour/Personalities

The following illustrates the three different images of the **BLACK** personality shown on the previous Colour/Personality Scale.

Michella's outlook on life is one of generosity. She is steeped in tradition and believes in sharing her time and expertise with anyone. Michella has created an environment of glamour in a small space, even though I have not noticed anything "small" in scale in her decor anywhere. Her living room seatings are traditional in style and covered in off-white damask, enhanced by an array of exquisite pillows. Her coffee and sidetables are piled high with large books on art, and intermingle with handsome antique silver candleholders and crystal bowls of potpourri. The walls are covered with original art from floor to ceiling. A down-to-the-floor damask tablecloth, tinted with tea for a mellow, antique look, graces the round dining table. The centerpiece is an enormous dish filled with tea roses. This is the environment of a **BLACK** personality who, rooted in tradition, chooses **OFF-WHITE** rather than crisp **WHITE** as her Accent Colour. She herself is a striking **BLACK** personality, looks great in **BLACK**—all five feet four inches of her.

Marcia's living room furniture and accessories are a collection of things found on her travels with her husband. The large corner seating arrangement is covered in stone-coloured suede and rests on a pink marble floor enhanced with a scattering of tribal rugs. Two black leather and chrome chairs complete this grouping. The tables are made of brass and glass, and are decorated with candleholders bearing eighteen-inch tapered, pastel candles and three large ice buckets filled with pink orchids. Several pillows covered in tapestry and embroidery which Marcia purchased in the Orient are stacked on the floor. A turquoise and pink Chinese kimono hangs over the fireplace and most of the other walls are filled with books.

Her dining room decor is equally dramatic. The walls are painted a high-gloss, deep **TAUPE**. Two huge palm trees stand on the pink marble floor, and the antique mahogany table is set with modern glass dishes and glasses. The rooms have the distinct aura of originality created by a **BLACK** personality who has relaxed to the **TAUPE-STONE** colour level. These surroundings are not for the timid. Fortunately, there's not a timid bone in Marcia. She enjoys her unique surroundings, her lifestyle and her collection of black clothes.

The foregoing demonstrates clearly that the basic Parent Colours influence one's choice of environment even if one's colour scheme is **BLACK**. Michella relates to **NAVY** personality traits and Marcia to the **TAUPE** or **COOL-BROWN** image, which is evident in their choice of decor.

June is my third example of the **BLACK** personality. She relates to the **OFF-WHITE** and warm **BEIGE** of the **BROWN** image, which is apparent in her warm, relaxed personality and the comfortable atmosphere of her home. She must have had fun creating her own beautiful environment with pieces purchased in her travels or from her own decorator store, and adding her own sculptures and those of well-known artists such as the large wall mural pictured in her dining room. The wide-plank oak flooring, oversized furniture and the large fireplace give her home a lovely ambiance, as seen in the photo. She created the exquisite table setting for the camera in the same manner she always does when she entertains her family and friends. At Thanksgiving, Christmas and other holidays her home is always filled with her family, children and grandchildren. She and her partner do the cooking in their streamlined stainless steel kitchen.

I sincerely enjoy interacting with these creative women who are not intimidated by large-scale furniture, exotic accessories and their all-encompassing lifestyles. They are truly the personalities who wear **BLACK** well as their costume, without intimidating others. These women can enter a room full of people unescorted and exude self-assuredness.

These examples of the **BLACK** environment demonstrate that the outgoing **BLACK** personalities all have a style of their own according to the traits of the **NAVY**, **COOL-BROWN** and **BROWN** Parent Colours. Interestingly, this information translates to their style of clothing as well.

The Black Clothing

The personality who is drawn to the non-colour **BLACK** is avante-garde, and a modern and fashionable wardrobe is appropriate. The **BLACK** personality influenced by her **NAVY** background who is rooted in tradition, however, will wear elegant classic clothes. The least constructed garments are mostly worn by the non-conformists, **COOL-BROWN**. Most **BLACK** personalities prefer garments made of good quality fabrics and workmanship with quality accessories.

It is a tall order to fit into the costume of the **BLACK** personality. **BLACK** is the most misunderstood colour in the Basic Colour group because so little consideration is given to the image it portrays. Almost everyone has **BLACK** in the closet or has used it as a major part of her wardrobe at some point in her life for several reasons. **BLACK** is practical, it makes one's figure appear slim, everything goes with it, and it is easily available in a variety of styles. But only the self-assured wear the non-colour well.

At the time of this writing, **BLACK** is a fashion colour and is worn without any consideration for what it conveys.

Advice for a basic wardrobe in **BLACK**: Please examine the colour personality scale of the different personality traits first. Is your nature more traditional? You will most likely choose clothes that hug your body. If you are a non-conformists you will probably choose softer fabrics and style, and if you practice the outgoing "open-door-policy." You will most likely prefer clothes with volume.

A complete wardrobe plan follows in the Blueprint section, but here is an example: When you are looking for a jacket to go with all the skirts, pants and dresses in your wardrobe, let it be one of current style (Do some research). It is the **BLACK** (and **RED**) personality who inspires others with her style—so go for it! The elegant jacket is usually fingertip length, not structured, made from a fabric that moves. Skirts may be narrow and worn at fashionable lengths. Gored or draped skirts work better at a longer length. Choose the most glamorous blouses (not shirts) in silk in your signature colours, i.e., **WHITE**, **OYSTER** or

OFF-WHITE, add a black cashmere turtleneck sweater for one occasion, and perhaps a draped, beaded one for another. Buy your shoes, handbags and belts in the best quality you can afford. The picture shows a designer seven-eighths jacket and dress in wool crepe. You see the unconstructed style—no buttons—and the choice of accessories which suits this self assured personality. The coat also makes a fashion statement with the wide banding of heavy satin, which matches a pair of satin pants (not shown). The black personalities have no challenge creating a costume for themselves and wearing beautiful oversized shawls or stoles to make a grand entrance.

Colour + Texture + Style = Personality

BLACK + Crepe + Grand = Dramatic

e

Gems from my Collection

I met my good friend Samaya many years ago when she worked in the world of finance and wore **BLACK**. She changed her profession to become an interior decorator and then returned to the busy corporate world. During these changes she re-decorated her apartment so that it would provide a soothing environment. Over the years she has used fewer and fewer colours. Her apartment is now monochrome throughout—**TAUPE**, **STONE**, **OYSTER**. She has spray-painted her wicker furniture **TAUPE**. She decorates with large glass bowls full of potpourri, white roses or other flowers in pale colours. There are candles everywhere and an inviting array of damask pillows. She uses the same colours and decor throughout her home. You know that here lives a person who needs a place to rest from the outside world and its constant demands.

I was recently invited to re-do her wardrobe to better suit her current life. She borrowed a portable clothes rack which she assembled in the living room and which came in handy. I looked at those colours in her wardrobe which would take physical energy for her to wear: **BLACK** went onto the clothes rack first—lots of it, followed by **HOT PINK** and **TURQUOISE** and finally **BOTTLE GREEN** and some **PURPLE**. As I arranged the clothes on the rack according to colour, quite a pretty wardrobe emerged for someone, but not for Samaya. She is a **TAUPE** personality. She likes soft feminine clothing. She prefers to wear long silk skirts, lace tops, chiffon scarves and silk robes—sitting on her couch sipping champagne. But that is not what her life is about right now. She needs clothes for the office. We planned her wardrobe around colours which suit her personality and surroundings in a style appropriate for the business world. **COOL-BROWN**, **TAUPE**, **STONE** and **OYSTER** are her Basic Colours which she likes to brighten with **SAGE GREEN**, **DUSTY PINK** and **PALE DUSTY BLUE** (the same colours are in her oriental carpet!) As you can see, for a **TAUPE** personality "bright" translates into pale colours—not **RED**. We chose versatile pieces for her wardrobe so that each piece can stand alone or become part of another image.

(You will learn more about using pieces of your clothing like accessories in the Blueprint section.) We chose two suits, one in **TAUPE** and one in **STONE**—both came with a long skirt and pants. We added silk skirts, silk blouses, lacy tops and sweaters in her accessory colours.

Samaya's closet is now a reflection of her apartment and, of course, suited to her personality and lifestyle. Oh, yes, a friend of hers bought all the clothes on the rack in the living room. How lucky can one be?

Black Style Change

Maureen consulted me about the look she should portray. She was a successful hairdresser who owned several salons and her business was expanding. There is a fine line between being appropriately dressed for business while expressing your creative talents. Maureen thought that her unconstructed (**BLACK**) costume should be acceptable. I believed that if she continued to dress in **BLACK** unconventional styles, she would convey that she either had a silent partner who ran the business, or that she was merely on the payroll. On the other hand, if she dressed like a successful entrepreneur and business person in a black suit, her clientele would probably think she couldn't be creative enough to style hair.

To choose the attire which conveys success in a creative field one must always select quality. The colour **BLACK**, which seems to be the one that most creative persons are drawn to for its inherent drama, can stay. The style should be modern and fashionable but the quality of fabrics and accessories make the statement that demonstrates financial success.

Maureen followed this advice and made it her personal statement. As soon as she entered one of her salons, her presence immediately conveyed the fact that she was the owner. It was great for her self esteem. This kind of transformation can be applied to other creative occupations as well.

Black to Off-White

Dark-haired and beautiful, Judith was one of my clients who wore **BLACK** with all the bright Accent Colours to suit her vibrant personality. She is her own boss and manages her busy retail clothing store. One day she found herself without a husband. The divorce shattered her self confidence and left her bitter about life in general and men in particular. Her black clothes made her look quite angry. It is important to know that **BLACK** as a colour also expresses darkness, hostility, anger and confusion. This is why I believe that **BLACK**, in order to convey a positive statement, should only be worn by people who have ample energy to overcome the negative effects of **BLACK** and make it one of positive action. I suggested a change to amply-cut, soft fabric, **OFF-WHITE** clothing and used her bright colours for accenting, e.g. a large red shawl or red shoes.

The combination of pale Basic Colours and the softness of fabric made her look mellow, yet secure at the same time. She wore a softly tailored red suit with bone accessories at the last appointment with me and showed off her new wedding band. **BLACK** was never included in her wardrobe again.

Black to Silver

Coleen was introduced to me by her friends who had given her the gift of a private consultation with me. Since she had been ill for a long time, I met her at her home. She wore black slacks and a bright red jacket. She is not an outgoing person and her illness made her tired most of the time. You probably guessed it by now. I changed her strong, active colours—which were an effort for her to wear—to soft tones of **SILVER** and pastels in order to comfort her during her recuperation period.

Black to Navy

Jim is another client of mine. He is a well-known public speaker. His uplifting personality has encouraged and inspired many who listened to his speeches. Some people found him overpowering and intimidating, especially when he wore a burgundy jacket, black pants, white shirt and a red striped tie. He needed a calmer look for selling and promoting his ideas.

Although the colour **NAVY** suited his profession (having made his natural gift for promoting ideas marketable) we decided on using **GREY** for his "stage life" and **NAVY** for his business attire. **BURGUNDY** remained as a favoured shade for accessories, ties, sweaters and even pants. He told me that the grey suit made him feel comfortable and "one" with the audience. He was convinced that people now approached him with much more openness.

The Pale Basic Personalities

THE **PALE BASIC** PERSONALITIES PLAY THE WHOLE SPECTRUM OF colours against the pale canvas of their environment and their clothing.

Colour/Personality Scale

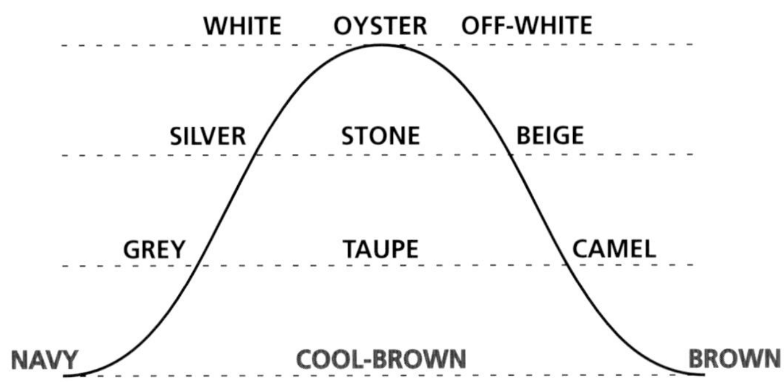

A separate scale is used for the **PALE BASIC** personalities because of their relationship to the three Parent personality traits (just like for the black personalities).

- **GREY–SILVER–WHITE** leans toward **NAVY.**
- **TAUPE–STONE–OYSTER** associates with the **COOL-BROWN** personality traits.
- **CAMEL–BEIGE–OFF-WHITE** relates to **BROWN.**

Each group of colours arranged vertically represents the three Pale Basic colour personalities. Each group of colours arranged horizontally represents the three levels of intensity of personality traits. The grey, taupe, and camel are the most pronounced for the Pale Basics.

Grey–Taupe–Camel

As previously mentioned, moving upwards on the vertical axis of the Colour/Personality Scale diagram, the colour (and personality) becomes lighter. These Pale Basics convey a "lighter attitude" than the one associated with their darker Parents. Communication generally becomes more open and people tend to find it easier to relate to the lighter colours. The darker Basic Colours can create distance. I have found that personalities drawn to the lighter shades are comprised of large enough numbers to warrant designating **PALE BASICS** as distinct colour personalities. **NAVY** moves up to the lighter Basic Colour **GREY**, **COOL-BROWN** moves to **TAUPE** and the **BROWN** personality moves up to **CAMEL**.

Silver–Stone–Beige

The next step up the scale reflects the personality that still makes the statement of the Parent Colour but psychologically implies a greater degree of non-involvement. These colours provide a canvas on which the Basic Colour personality can be communicated by adding appropriate Accent Colours.

White–Oyster–Off-white

The lightest shades of the basic Parent Colours are at the top step of the scale. These colours denote relief from stress, and take the least energy to wear or to use for surroundings. These colours create a calming effect and provide a good background for a vibrant personality. **WHITE**, the lightest shade of the cool colour family, communicates crispness and preciseness of personality.

People who are drawn to these lighter shades of the **PALE BASICS** have, psychologically speaking, accepted their basic roots, regard them as valid in their lives and are secure enough to be open to new experiences. The trying out of the new and sometimes unconventional (according to their Parent Colour personality) manifests itself in all areas of their lives. What does it mean, "having accepted their Parent Colour personality or basic roots?" When you know what each colour personality portrays, then you are free to accept your chosen colours as valid in your life, or you can change your attitude and clothes in Colour, Texture and Style to communicate a different message.

During my teaching years I frequently chose the **PALE BASIC** colours for my clients. I wanted to make an impact on the environment at large and bring in light. By working with the **PALE BASIC** colour scheme I gave my students the opportunity to use Accent Colours, different textures and styles to communicate a different message.

My preference is to work with the ***PALE BASIC*** *colours in my environment and I have used* ***BEIGE*** *frequently in my wardrobe during my years of teaching and consulting.* ***BEIGE*** *and* ***OFF-WHITE*** *make me feel most comfortable because they do not take physical energy to wear.* ***BEIGE*** *conveys calmness and is non-threatening to people to whom you are conveying your ideas. When I wear* ***BEIGE*** *or* ***OFF-WHITE****, I do not accessorize with other colours.*

On the other side of the coin, the strong action colours ***RED*** *and* ***NAVY*** *are in my colour scheme. When I wear* ***NAVY****, I am aware that I can create distance and when I wear* ***OFF-WHITE*** *I get a lot of hugs. (Try it!). Because* ***RED*** *is my true colour, I wear it with assurance and confidence and receive many compliments. Yes, I like a variety of colours, but I would not venture outside my Accent Colours related to my Basic Colours.*

When you first start to experiment with colours as a statement, rather than a pigment, it might seem strange to you, but I assure you that the colours have meaning—every one of them. Here are some examples:

A phrase frequently heard from my students directed at their friends is: "Dont't be so **NAVY**"—meaning don't be so conservative. It could also mean: "Open the first button on your shirt or blouse." A **NAVY** student proclaimed: "I had a really **BROWN** weekend. I did not make the beds." One of my students (**BROWN**) announced that she suffered through a **NAVY** Sunday. She cleaned house and polished the siver candle sticks, which she wanted to use on her Sunday dinner table. An efficient **BLACK** personality was overheard saying: "I had a **TAUPE** weekend. I read three books on poetry and burned candles galore." See!

The following illustrates the three different images of the **PALE BASIC** personalities which are charted on page 60.

THE PALE BASICS OF THE NAVY PARENT

ELEANOR, MY CLOSE FRIEND AND BUSINESS ASSOCIATE FOR MANY years, provides a good example of how to communicate her personality by her choice of the colours **GREY**, and **SILVER**. She owns a complete wardrobe in those shades in textured and soft fabrics (no lapel on jackets—no navy blazer) which she accents with primary colours. She tops an outfit with an unconstructed red coat—not a trench coat. The whole look portrays her as the youthful, action-oriented person she is. Even though her background is **NAVY**, her personality is often **BROWN**. Her house is full of family and friends most of the time. She and her husband Peter, love the outdoors and enjoy hiking. The combination of the **NAVY** personality and **BROWN** lifestyle made the **GREY** and unconstructed style the perfect choice for her wardrobe as well as for surroundings. Their house has character. Each room displays their signature with the use of Danish modern furniture to which Eleanor adds bright red-and-blue-textured fabric for pillows. Paintings, soapstone sculptures and beautiful pottery pieces filled with bright coloured flowers enhance the inviting atmosphere of their home. The outside of their house is painted grey with natural wood trim and is brightened with a large basket of yellow daisies at the front door.

The Pale Basics of the Cool-Brown Parent

My daughter Fiona is a typical **TAUPE** personality with a touch of **NAVY** preciseness. **TAUPE**, **STONE** and **OYSTER** are her signature in garments that she wears in unconstructed styles. Because of her international business activities, she always anchors that look with a structured jacket or coat. In the summer she uses **NAVY** and lighter shades of **BLUE**. Her style is avante-garde and her personality trait is unconventional. The combination of the two gives her clothing a distinct style which is uniquely hers. Fiona's environment includes a collection of art and artifacts from foreign countries displayed on open shelves against taupe walls. The floors are covered with Oriental carpets and her comfortable corner couch is strewn with pillows in soft colours. Her glass-topped dining table is set with a collection of fine china and unusual glasses. She has flowers and candles everywhere. This is another true **TAUPE** environment, original and hard to duplicate. She is in harmony with her surroundings both in Colour and Style inside her home as well as in her outdoor space.

The Pale Basics of the Brown Parent

The third **PALE BASIC** personality relates to the **BROWN** Parent Colour. Like all **PALE BASIC** personalities they take liberties with the texture and style of their clothing and environment. They mix and match but never lose their **PALE BASIC** colour signature. The following example illustrates another typical **PALE BASIC** personality.

Roberta was a client of mine for many years who truly enjoyed wearing these pale, warm shades. She decided to decorate their West Coast design home in the same colour scheme. **OFF-WHITE** hand-woven fabrics for the over-sized Chesterfields complimented the **CAMEL** coloured leather armchairs. Lots of off-white crewel embroidered pillows filled a long window seat. She changed the pillow covers in the winter with her own hand-woven ones in her warm accent colours of **TOMATO RED**, **GOLDEN YELLOW**, **WARM GREEN** and **PERIWINKLE BLUE**. She also displayed balls of wool of these colours in a large basket by the fireplace.

Roberta mixed stoneware dishes with crystal on her long oak dining table which was usually shared with many friends. She definately had an open-door policy but with a bit of **NAVY** influence.

Her clothes showed the same friendly outgoing personality traits but with a mix of polish. For example: she wore corduroy pants with silk blouses, or a hand-knit heavy silk sweater with garbadine pants—all in **BEIGE** or **OFF-WHITE**, or a **CAMEL** coloured tweed jacket with a long beige twill skirt and high ankle boots and a slouch hat.

Roberta used a lot of **RED** accents. She wore her over-sized red cashmere shawl with everything as her only accessory for her otherwise **BEIGE** costumes. She used her **PALE BASIC** signature perfectly.

Gems from my Collection

What happens when two people with different colour personalities live together? Fortunatly there are colours which are suited to both. When you and your partner have decided on your individual colour schemes, look for a common denominator either from the Basic or Accent Colours. Realize that all colours can be used in all shadings and depths. Let me use our colour scheme and environment as an example:

My husbands colours are:	*Basic*	*Accent*
	NAVY	***RED***
	BEIGE	***GRASS GREEN***
	OFF-WHITE	***YELLOW***
My colours are:	*Basic*	*Accent*
	CAMEL	***RED***
	BEIGE	***OLIVE GREEN***
	OFF-WHITE	***NAVY!***

You can see our different personalities, can't you? My husbands **NAVY** *characteristics are softened by the* **BEIGE** *colour and what it represents, and I have a lot of* **NAVY** *traits. My husband, says his* **NAVY** *traits are built in, mine are aquired. We choose* **BEIGE**, **BLUE** *and* **RED** *as the common denominator for our home.*

Black and Beige

Most of my clients whose houses I visited were women who decorated their house by themselves along with a few suggestions from their husbands. Some used the services of an interior decorator. The husband's domain, the den, is often the only masculine-looking room in the house. John is the only man I know who took charge of decorating the whole apartment which resulted in reflecting the strong signature of his personality (**BLACK**). I am sure his wife was consulted and her soft image—she is a **BEIGE** personality—showed up in the accessories.

John is a truly action-oriented, high energy businessman who bought a new apartment overlooking the downtown financial district. He immediately transformed every square inch of a very ordinary apartment into a truly handsome showcase. He installed new stainless steel kitchen appliances, granite counter tops, used black iron curtain rods with white cotton curtains and added tall black iron candle holders to the decor. He rough-plastered the walls, replaced the wall-to-wall carpet with terra cotta tile, bought black leather seating arrangements and, of course, a state-of-the-art sound and video system. He wears **BLACK** and drives a black car. His wife, Karen—tall and beautiful—is an excellent cook (and so is he); and she creates dried flower arrangements for their apartment. The scheme works for both of them.

Taupe and Navy

A **TAUPE** client of mine was concerned that her new **NAVY** partner would ruin the look of her living room by adding a special chair for himself. The **TAUPE** personality would have to unclutter the living room and give it some structure in order for a **NAVY** personality to share it. A chair or two instead of cushions on the floor, and a reading lamp to replace the candles, would result in the kind of room that both of them could enjoy.

Brown and Navy

Doris recently married a **NAVY** personality who is drawn to the traditional navy attire which indeed suits him best. It is really true that **NAVY** personalities like the style of dress associated with that image and prefer the colour **BLUE** for their clothing. I suggested some **NAVY** accessories for their home, a small crystal chandelier, a pair of silver candlesticks, a blue velvet pillow and maybe even some blue flowers on the deck which is a continuation of their beige and apricot living room. It takes so little to address and satisfy both personalities who live under the same roof.

Navy and Beige

*When we had two homes, one in town and one in the country, I had my office and studio in the townhouse. I wore only **BEIGE** and **OFF-WHITE** and our townhouse was* *decorated in the same colours. No colour accents! All* **BEIGE**, **OFF-WHITE**, *glass and brass. Period! Our island home was different—very different. But I would like to tell you about our present home back in the city. We renovated every room, floors, walls, ceilings, fireplaces and kitchen. When it was completed, just about every wall in the house was painted a* **DARK BEIGE** *with antique white moldings and trim. The floors were finished with rustic beige Italian tiles. The floor-to-ceiling drapes and sheers are handwoven, unbleached cotton. We recovered the Chesterfields in off-white cotton. Only my husband's office is* **PERIWINKLE BLUE** *with white trim.*

A Basic Colour scheme, whether it is in your closet or home, makes life a lot easier. Using one or two Basic Colours which suit both personalities makes both feel relaxed and in harmony with their surroundings. Use your Basic Colour scheme for background and play the Accent Colours against it. You would probably not recognize our home now. It is full of colour, warm red and blue prints, lots of blue hand-painted dishes, blue table linens, and yes, I do grow blue flowers for the man in my life. My husband likes all colours as long as they are **BLUE**. *My studio upstairs has lots of windows and the high walls are* **PALE BEIGE**. *(I am constantly working with colours both with my hands and with my mind.)*

Colour, Texture and Style Equals Personality

In the foregoing chapters I have explained the different colour personalities which are related to personality traits. Knowing this, the choice of a Basic Colour as a pigment becomes secondary to the selection of Texture and Style. Let me summarize the basic messages communicated through Colour, Texture and Style of your garments. Your clothes will show that you are a person, not a people.

To convey a precise and efficient message, you will choose the true **NAVY** image which conveys a crisp, traditional, precise and efficient look. The darker the colour and the more angles in style the more distance the image creates.

Following is the **NAVY** image in decreasing order of its impact using different colours for a jacket.

1. Colour : **BLACK**
   Texture: worsted
   Style: close to the body
2. Colour: **NAVY**
   Texture: worsted
   Style: close to the body
3. Colour: **COOL-BROWN**
   Texture: worsted
   Style: close to the body
4. Colour: **BROWN**
   Texture: worsted
   Style: close to the body

BLACK as a Colour, worsted fabric as Texture, and close to the body Styling presents quite a formidable image. It requires a considerable outlay of energy to engage in a meaningful conversation with that person, and it is difficult to get to the heart of things. When you are wearing **BLACK**, you will spend extra energy to convey a warm personality. It is interesting to know that the true outgoing **BLACK** personality has no trouble doing just that—no effort at all. Needless to say, comunication gets easier the lighter or warmer the colour or the softer the texture.

The opposite of the distance creating **BLACK** image is the relaxed look of the **BROWN** personality.

Following is the **BROWN** image in descending order of conveying a relaxed message using different colours, again for a jacket.

1. Colour: **BROWN**
   Texture: tweed
   Style: easy fitting
2. Colour: **COOL-BROWN**
   Texture: chenille
   Style: easy fitting
3. Colour: **NAVY**
   Texture: knit
   Style: easy fitting
4. Colour: **BLACK**
   Texture: velvet
   Style: easy fitting

Not everyone will feel good wearing a tweed jacket, true! But the **BROWN** personality will.

Knowing what your clothes communicate makes it easier to convey the desired message and it will show your "I-care" attitude.

When you look at your Accent Colours in the next chapter, carry with you the thought of Texture and Style and what they will communicate.

For example, take the look of a classic cotton shirt versus a soft sweater. The choice is yours.

Accent Colours

The accent colours you wear, once again, reveal your personality. In this chapter, I will guide you so that you can use them effectively to complete the visual impact of the personality which you wish to communicate to others. Although you will use these colours as accents, they can also make a statement on their own, if desired.

TRUE RED (and related shadings) is the most important of all Accent Colours. **TRUE RED** is intense, vital and stimulating, and is favoured by the young at heart of all ages. It can be exciting without being threatening. The person who likes **RED** has a variety of interests, is optimistic, strives for a life rich in activity and enjoys people. Most everyone can relate to **TRUE RED** but only the **TRUE RED** personality will wear it to make a statement. If you want to be swept away with the idea of traveling on the Concorde or the Orient Express to Paris or Venice, find a **TRUE RED** personality to start you dreaming, and you might even discover that leaving the familiar can be fun. **TRUE RED IS AN ACTION COLOUR.**

The other shades of **RED** suitable for different personalities, and all the other glorious accent colours are explained in the following pages.

Accents for the Navy/Grey Personality

As mentioned in the chapter on basic colour personalities, **NAVY** makes a definite statement about being orderly, precise and action-oriented. I would like to explain once more that by taking a step up on the Colour/Personality Scale, the depth of colour becomes less intense as do the personality traits. I have observed that a large number of people, perhaps because of the nature of our culture, have some amount of **BLUE** in their personality, even if it is only **PALE BLUE**.

RASPBERRY a colour in the **COOL RED** family, is also an action colour. It conveys brightness and an outgoing personality.

HOT PINK is the next shade in the **COOL RED** family. It is another action colour and is often used with **SILVER GREY** to make a vivid statement. This colour is for the person who wants to be seen, and is not intended for the timid.

CORNFLOWER BLUE is a bright primary colour which relates to happy outgoing activities. Children love it and so do people who are fun-loving and young at heart.

PALE BLUE can be added by the action-oriented **NAVY** personality in order to soften their appearance. In contrast, **PALE BLUE** as an accessory colour for the relaxed **BROWN** person, conveys order and conservatism and, in fact, it sharpens their image.

YELLOW/LEMON (**YELLOW** with cool undertones) is another active, almost electric colour. It stimulates like **RED**, but to a lesser degree.

All the above colours are wonderful accents for your home decor and also for flower arrangements.

Accents for the Cool-Brown/Taupe Personality

COOL-BROWN/TAUPE PERSONALITIES HAVE A NATURAL ATTRACTION to "off" colours (i.e., colours softened by adding other colours such as grey) which portray their interest in aesthetics and their preference for originality.

MAUVE, the colour of lavender and the haze surrounding a sunrise, evokes a feeling of magic to the sensitive **TAUPE** personalities. Mauve is for the person who collects butterfly wings.

DUSTY PINK (**PINK** with **GREY** undertones) is in the same category as **MAUVE**. These personalities are sensitive and have a special need for compassionate understanding. They delight in their magical relationships with others.

SAGE GREEN is another colour which signifies physical ease in clothing and surroundings.

FRENCH BLUE makes a more subtle statement to portray **NAVY** personality traits for the **TAUPE** image.

Collect candles in all heights in the above colours. How about cushions in different textures for different seasons? Fill your balcony, terrace and home with these wonderful shades of flowers to set a meditative mood.

Accents for the Brown/Camel Personality

GARNET is a warm RED. The impulsive vivacity of TRUE RED can be reduced by adding brown tones, thereby rendering it more peaceful. **GARNET** better suits the **BROWN** to **BEIGE** or **TAUPE** personalities. It is less assertive than **TRUE RED** and therefore easier to wear allowing communication with others on a more relaxed basis. **GARNET** is an appropriate colour for creating calm surroundings.

TOMATO RED is a warm **RED** that also possesses brown undertones. People who are attracted to this colour like contact with people, and prefer the comfort of relaxed surroundings.

OLIVE is **GREEN** with warm undertones. It is a calming colour that has some of the characteristics traits of a **BROWN** personality. The person attracted to **OLIVE** is flexible, enjoys the quietness of nature and loves all creatures. These people are often relaxed in their attire and belong to the **BROWN** and **BEIGE** personality group. Some people see **GREEN** as a symbol of being environmentally conscious, which is another twist on colour as an expressive tool. A soft pale shade of **GREEN** is frequently used for hospital interiors for its calming effect.

YELLOW (with warm undertones) is a light and cheerful colour. Symbolically, it represents the light and warmth of the sun. **GOLDEN YELLOW** expresses emotional content, a cheerful spirit and conveys a feeling of happiness. People are attracted to **GOLDEN YELLOW** for these very reasons. This happy, sunny shade of **YELLOW** often makes its way into kitchens.

PALE BLUE, as an accessory colour for the relaxed **BROWN** person, conveys order and conservatism and, in fact, sharpens their image.

These accent colours just ask to be added to a quilt or handwoven shawl. Fill baskets with flowers for your livingroom and bright yellow candles and napkins for your dining room table.

Accents for the Black Personality

I have chosen lovely velvets in the picture to illustrate the glamour and drama of the **BLACK** personality. Silk charmeuse, taffeta and black lace could have portrayed the image of a typical **BLACK** outgoing, self-assured personality. These characteristics are of importance in order to wear **BLACK** and the following Accent Colours well.

BURGUNDY is the most action-oriented of the colours derived from **RED**. Most **BLACK** personalities relate to it. Some **BLACK** personalities will wear a complete burgundy outfit, while others find small touches of it enough. The darker the shade of **BURGUNDY** the more physical energy is required to wear the colour, and the more aggravating it can be to others. People who wear **BURGUNDY** (other than the true **BLACK** personality) will probably need a vacation more often.

EMERALD GREEN with cool undertones is an action colour that signifies the **BLUE** characteristics of precision, order, ambition and authority. For that reason **NAVY**, **BLACK** and sometimes **GREY** personalities are attracted to it.

PURPLE is a hue which seems to have an aesthetic appeal for people who are interested in the arts or have strong religious feelings. **PURPLE** can also cast a gloomy impression on the people wearing it. Many shades from the deep **AUBERGINE** to the lighter **PERIWINKLE BLUE**, are easier to accept than **PURPLE**. The colour **AUBERGINE** gives most personalities a more reflective appearance; and a whole room painted in **PERIWINKLE BLUE** adds calm to the environment.

I have seen many of my **BLACK** personality friends use these dramatic accent colours in their decor, e.g., large gold framed paintings with wide emerald green matting, burgundy silk wall hangings over the fireplace, or oversized purple pillar candles on the dining room table. None of these dramatic colours are intimidating to the **BLACK** personality.

Accents for all Personalities

Add these to your Basic Colours for changing your image, or to portray a certain mood in your attire and surroundings.

SHELL PINK conveys the need for peace and quiet. It suggests complacency and for that reason an older person might be drawn to it. **SHELL PINK** would not be my choice for a jogging suit.

APRICOT makes a calm statement and is a colour for people who like to feel pampered. It is considered sensuous.

BLUSH, a lighter shade in the same family of **WARM REDS**, is conducive to calming the nerves and relaxing the body. It is chosen for intimate apparel and/or surroundings.

PERIWINKLE BLUE is favoured as an accent by the **CAMEL** or **BEIGE** personality to convey a **NAVY** trait, but with less intensity. Along with their Basic Colours it makes a more definite business statement without losing the individual's personal signature.

I believe very strongly that the majority of people find use for any of the palest colours in their wardrobe and/or surroundings. The pale shades of the Basic and Accent Colours soften the image portrayed. For ease of living, increased calmness and feelings of serenity, the **SOFT PASTELS** and the **PALE BASICS** are the answer.

Continue to the next chapter to learn how you can develop "Your Signature"—even using your Accent Colours.

Creating Your Signature

The information in this book is intended to guide you to look like yourself, not like anybody else. You have your own style, your own signature. You will also "know" why you are wearing certain Colours, Textures, Styles and accessories.

Colour is definitely a signature and when you make it yours, you will know that you can pack a suitcase and go on a trip around the world with a colour-coordinated wardrobe without feeling you have left dozens of colours behind. I have had students tell me at their first seminar with me that they like all colours and wear them at random, and each room in their home was decorated in a different colour. I used to say that I would not like to travel with them, because they would have too many suitcases. They are the kind of people whose closets are full of clothes but they have nothing to wear.

Make your colours your signature, **NAVY**, **COOL-BROWN**, **BROWN**, **BLACK** or the **PALE BASICS**; choose one. You will be glad you did.

Do you always like to wear **BLACK**? Then make it your signature—you now know what image you will portray. How about **PINK**? My friend Alannah who also was my client and a student for years is known for the colour **PINK**, from pale to hot. Pink is her signature. Come to think of it, all my friends have their own colour signature. It is second nature to them, as it is for me.

Texture in your clothing can be a signature as well. My friend Michella collects beautiful charmeuse blouses in all colours. Another client of mine buys handwoven shawls wherever she goes. She collects them in wool, silk and chenille and wears, carries or drags around one of her finds. That is one of her signatures.

Let's look at the style of each personality which could become your signature. The style of clothing as the signature for the **NAVY** personality are garments that are worn close to the body, at least at one point. They are well constructed and made of quality fabrics. The **COOL-BROWN** image can be made up of a collection of ethnic and antique clothing. **BLACK** attire is usually carried out in a slim or softly draped silhouette. The **PALE BASIC** style is all of the above—one of each, e.g, a long slim skirt in off-white wool with an antique lace blouse worn with a crushed leather belt at the waist and a red suede jacket. You may not find that style in every store, but you have discovered your own signature and style.

How about shoes? You know that your whole costume starts with them. Are you collecting shoes and making them your signature? My friend Jean who is seventy-nine years young always wears high-heels, even when she visited us on the island to cut her Christmas tree. I have a picture of her cutting her tree in high-heels and a white mink coat. Her shoes are her signature, she is known for them. My friends and students tell me they think of me when they see gold shoes. I have them in all styles. That is one of my signatures.

How about your accessories? Accessories for the **NAVY** or classic look consist of strands of pearls in all lengths and sizes, gold and silver chain necklaces, diamond, sapphire, emerald and ruby rings and pins in traditional settings, narrow gold and silver bracelets and small watches, narrow leather belts with good buckles, a silk scarf and plain pumps. The **COOL-BROWN** collects her accessories. She most likely owns an antique watch on a gold chain, cameo pins and earrings, a garnet brooch, rings and earrings, turquoise and silver jewellery, antique lace scarves, unique sandals and boots. Accessories for the **BROWN** image include, wide bracelets in wood or jade, generous handwoven shawls, sturdy boots and sandals. The elegant **BLACK** accessories are usually generous in size: diamonds, large pearls, bracelets, several rings, earrings, crushed leather belts, large silk and chiffon scarves and high-heeled shoes.

The **PALE BASIC** is a mix of all of the above. These individuals most likely hang an antique brooch on their opera length pearls or mix diamond rings with an ivory or jade band. They wear a turquoise and silver belt with classic pants and a cashmere sweater, and may top the outfit with a large paisley shawl. Would you not wear this costume to portray your unique personality now that you know and feel so right playing your accessories against your **PALE BASIC** canvas?

Your Signature in your Environment

One of the results of selecting Colours, Textures and Style for your clothing in harmony with your personality is that you will also become more aware of the colours in your surroundings. When you use your colours and your special touches in interior decor, the room will reveal all kinds of things about yourself. Your true personality will show by the way you accessorize and set the table or arrange flowers. Each room will have your personal signature by incorporating appropriate Colour, Texture and Style. Accessories are the easiest and least costly means to use in order to create an environment suited to different personalities, especially if they live under the same roof.

Your Signature for your Living Room

Your living room can reflect your personality through the use of pillows and area rugs in different Colours, Textures and Styles. Area rugs can readily change the character of your living room. Special containers for your favourite flowers are another signature; so are lampshades and candles—how many, what colour and where you place them. Picture frames sometimes say more than the pictures. All are elements that can be used to enhance your living room and personalize it with your special signature.

COLOUR	LAMPSHADES	CANDLES	PICTURE FRAMES
NAVY	Pleated silk is appropriate for a painted porcelain lamp	White & pastel tapers Silver holders	Antique silver, lots of them
TAUPE	Stained glass shade on an art nouveau sculpture	All colours All types	Velvet frames for small art reproductions
BROWN	Textured, linen shades suit a milkcan base	Beeswax candles ceramic holder	Oak frames to hold all family pictures
BLACK	Gilded parchment for a black china base	Gold candles Large brass or iron candelabra	Plexiglass for a glamorous picture of you
PALE BASICS	Plain linen is the choice with ribbon trim to match your candles or decor	All colours, all sizes, in a variety of holders, anytime, rain or shine	Different styles and colours hold pictures which may cover a whole wall

COLOUR	PILLOWS	AREA RUGS	FLOWERS & CONTAINERS
NAVY	Silk, velvet, chintz in your favourite Accent Colours	East Indian, or Oriental to complement your colour scheme	Silver, china, crystal with roses and carnations
TAUPE	Ethnic embroidered covers collected in your travels	Tribal rugs, prayer rugs to continue your special signature	Sculptured glass in intricate design to show an unusual flower in shape or colour
BROWN	Corduroy, handwoven, tweed, fabric downfilled for extra comfort	Braided or hooked rugs for the "homey" feeling	Pottery, ceramic, baskets filled with wild flowers, grasses, dried bouquets and living plants
BLACK	Silk with gold cording and tassels, velvet, damask and all oversized	Fur or Oriental to create drama in your home	Large silver shapes for exotic longstemmed orchids, lilies, bird of paradise or silk flowers
PALE BASICS	Linen, cotton and all of the above	Your Parent Colour is often an indication of which rug to use and you may even hang it on the wall	A basket filled with roses, a silver teapot with daisies, a pewter pitcher with daffodils—the unexpected

Your Signature for your Dining Room

The dining room is another place that you can dress along with yourself. I sincerely hope that you use this room more often than at Christmas and Thanksgiving or when entertaining your friends. Remember your family comes first. Our 90s lifestyle has developed into an eat and run marathon. A bare kitchen table or counter with unbreakable dishes has become the lifestyle for many people, often forgoing the how-to of setting the table and how-to of using eating utensils.

Here are some short tips to refresh your etiquette and for those who think it is important to display an "I-care" attitude in all parts of their lives. It is so easy to say that "it does not matter," but pretty soon your whole life becomes an "it-does-not-matter" affair. Most of you know that forks are placed on the left side of the dinner plate. If a salad course is served first the salad fork is the outside utensil. The knife belongs on the right of the plate and the spoon sits next to it; dessert spoon and fork rest across the top of the plate. The spoon handle faces to the right, the fork handle to the left. The butter knife rests across the bread-and-butter plate, which is placed at the upper left corner of the dinner plate. It is that simple. Most of you also know that food is served from the left and drinks from the right. Remember, the little things in life do make a difference. Everyone can benefit from them. Have fun giving your tablesetting your personal touch. Choose table coverings, stemware and dishes that are a reflection of you. Use candles in colours that compliment your tablesetting. Make sure that they are at a height so that the flame is not at the eye level of your guests. For the same reason keep your flower arrangements at a reasonable height so that people don't have to play hide and seek among the gladiolas. Try not to select flowers for your dinner table that emit a strong scent as they will overpower the aroma of the more delicate food and desensitize your guests' tastebuds. Most of all, think of *yourself* in harmony with your surroundings and your personality.

COLOUR	TABLE COVERINGS	DISHES & FLATWARE	STEMWARE
NAVY	Damask and lace	China, often with pattern, Silver flatware	Cut or plain Crystal
TAUPE	Hand embroidered tablecloths and napkins	Delicate glass dishes or one of a kind antique Silver flatware	Collected pastel coloured glass & Crystal
BROWN	Handwoven placemats and napkins	Stoneware, pottery Stainless steel flatware	Glass or ceramic
BLACK	Mirrored placemats Damask napkins	Silver or Brass chargers with black or white square dishes Silver flatware	Modern design crystal or glass
PALE BASICS	Collects all to convey different moods for all occasions, formal or not	Plain white quality china plus stoneware in their Accent Colour Different flatware for different occasions	Plain style, uncut crystal glass, or other to set a mood

YOUR SIGNATURE FOR YOUR BEDROOM

THE BEDROOM IS THE MOST PERSONAL AND PRIVATE ROOM IN YOUR home and for that reason it tells a lot about you. Most of the bedrooms I have seen were feminine in flavor even if the rest of the house was not. Sometimes the bedroom displayed signs of masculine touches—less ruffles, the use of deeper colours or tailored bedspreads. (I had placed my husband's cowboy boots beside my bed as part of the decor for a long time.) Whatever your space, create a pleasing atmosphere by having things colour-coordinated. The sheets should blend with the tones of the bedspread and curtains. You would not choose blue and red-striped bedding for a pink-flowered room. Nor orange florals for a pale blue setting. When you shop for bedlinens and towels spend your money on the right colour or use plain white linens. Your personality is also on display by the accessories on your dresser or night table.

NAVY	A single rose and silver-framed pictures on the night table. An extra pillow or two on the bed.
TAUPE	Crystals hanging in your window. Antique perfume bottles and jars and fresh flowers on the night table.
BROWN	A wreath of dried flowers over your bed. Baskets with potpourri on the dresser.
BLACK	A decanter with sherry on a silver tray. Candles and crystal parfum bottles on the dresser.
PALE BASICS	All of the above with emphasis on Parent Colour and lots of candles.

Your Signature for your Guest Bedroom

The Guest Bedroom can also carry your personal touch, conveying that you care about your guests. If your guest uses a closet in which you store your off-seasons clothes, cover your garments with white opaque garment bags and provide adequate space and empty hangers for your guest. I like to place flowers and candles at the bedside along with an interesting book or current magazine. Everything announces a special welcome. I made some small decorative pillows filled with lavender for the bed, and added a basket full of scrumptious towels, soft facecloths and fragrant guest soaps. It is a good idea to have extra toothpaste, a toothbrush and a razor in the basket or in the guest bathroom. Don't forget a night light! A thermos or crystal decanter with orange juice and a glass on a tray adds another personal touch with which to greet your waking guests. I always enjoy these special treats when I am a guest at my friends' homes. Everything you do can bear your special mark.

Your Signature for your Bathroom

Your own bathroom becomes a wonderful place to give reign to your creativity. Check the chart how different personalities are drawn to specific colours and accessories.

NAVY	This personality likes an uncluttered and streamlined room in **WHITE**, **PALE BLUE** or **SILVER-GREY**. A space for everything (behind mirrored vanity doors) and everything in its place. Either bright-coloured towels are used for the action-oriented person or pale pastels to give the environment a relaxed feeling. Soaps in antique crystal bowls are displayed along with a vase of fresh carnations. Mirrored walls and good lighting add to the functionality of this bathroom.
TAUPE	A stained glass window, an art-nouveau-framed mirror or a leaded glass lampshade are some of the signature pieces for the **TAUPE** personality. A collection of crystal creampots is displayed on a table covered with antique lace. An Oriental or tribal rug serves as a bathmat.
BROWN	One wall is painted, or wall-papered with a country print in a dark colour in order to visually decrease the space making it cozy and suitable for a retreat. An antique free-standing bath tub adds to the atmospheres as does a handcrafted pottery wash basin. Several plants in baskets hang from the ceiling and a large basket on the floor holds lots of beige and rust coloured towels.
BLACK	One red wall and white or black fixtures with brass or gold plated handles set the scene for this dramatic bathroom. Glass shelves line one wall stacked with a generous supply of white towels and beautiful containers to hold creams, lotions and soaps.
PALE BASICS	Your signature is always the **PALE BASIC** colour used in the background to unify all else. The bathroom is no exception. Walls and fixtures are done in pale shades, and accessories add a splash of colour. Most people change the mood of their bathroom by simply changing the colours of their towels. Use a glass shelf to display collections, and let plants and fresh flowers surround personal treasures.

YOUR SIGNATURE FOR YOUR OFFICE

YOUR PERSONAL SURROUNDINGS EXTEND BEYOND YOUR PRIVATE home. Your place of work is a space which also bears your signature. If you have a private office or work space consider the colours you wear—they should not clash with the wallcoverings, i.e., wearing a bright red suit in an office with primarily burgundy and teal blue walls. Your aim is to be in harmony with your surroundings without foregoing your personal colour preferences.

The same should be a consideration for personnel in an elegant hotel or restaurant. Not long ago we stopped at a lovely old country inn for lunch. Everything about the interior and the park-like setting was executed to perfection. The dinning room was decorated in soft dusty pink and sage green velvet, the waitresses wore long green dresses and white lace aprons. The hostess wore an acid yellow jacket, purple blouse and black skirt! I am tempted to send them a copy of this book. Don't you agree?

Office accessories can have your signature as well: letter openers, business card holders, a pencil tray. (A coffee mug is not quite right for a professional office environment.) Pens and pencils are also made to convey your message. For your fountain pen try a different colour of ink to adopt as your signature. **RED** for action, **BLUE** for business or **BLACK** for preciseness, **GREEN** for friendliness and **BROWN** to identify you as an easily accessible personality. Your stationary could be done in your **PALE BASIC** colours, i.e., **BEIGE** or **OFF-WHITE**, **GREY** or **SILVER**, **OYSTER** or **WHITE**, using the darkest Basic Colour for ink such as **BROWN**, **NAVY** or **BLACK** for the lettering.

If you are self-employed be aware of the first impression your business card makes. Match it to your stationary. Do not make the print too small. Human development consultants or psychologists would be wise to choose **BEIGE** or **STONE**-coloured cards with soft lettering for their clientele. A card with powerful lettering and colour, e.g., **WHITE** with **BURGUNDY** is more suitable for people involved in sales. White cards with navy or black ink are effective for anything having to do with accounting and banking. Purchase your business card case, briefcase or handbag with your signature in mind. Use a regular briefcase for your business folders—not a duffelbag—otherwise your clients might think your dog slept on the financial report.

Empty your handbag or briefcase and see how the contents can be improved. Is everything colour-coordinated? Glass case, chequebook, makeup case, etc. These items do not need to be in the same colour but they do need to compliment each other. The following chart shows accessories for your desk.

NAVY	A silver framed photograph
TAUPE	A framed poem changed often
BROWN	A plant in a basket
BLACK	A budvase holding a single exotic flower
PALE BASICS	Everything organized in baskets or something acquired while traveling, a Murano glass paperweight, a small sandstone carving, always fresh flowers

Your Signature for Small Living Spaces

Any suitable living space, apartment, boat, recreational vehicle, summer camp or ski cabin can have an uncluttered look by using your colour scheme, and giving your private area your signature. If you are a boat owner you'll benefit from the following information. Sailing vessels and motor boats are sometimes decorated in non-nautical colours by the manufacturers. Colours such as **GOLD**, **ORANGE**, **GREEN** or **BURGUNDY** are not suitable for boat interiors. It is interesting to note that most boat accessories are available in the **BLUE**, **RED** and **WHITE**, traditional boating colours. Dishes, plastic glasses, paper napkins, nautical towels with emblems, as well as turtleneck sweaters, boating shoes, pants, scarves, toques and windbreakers are all available in the cool-action colours. When one is confined to the small space of a boat or any other small environment, it is even more important to be in harmony with the surroundings. If the colours of clothing and accessories become part of the decor it creates a visual effect of spaciousness, rather than clutter.

For example: you and your partner favour the **BLUE** colour schemes but your boat is carpeted in **BEIGE**, the seating arrangement is upholstered in bright **ORANGE** and the curtains are **GOLD**—how do you change the look to give it your signature? First, change the curtains to an off-white cotton. Add blue cotton pillows in all shades to the couch—ignore the wild orange colour until somebody spills some red wine. Blue sheets, blankets and towels will make the space look more as though it belongs to you.

If your colours on the other hand are **BEIGE** and **BROWN**, acquire nautical-looking off-white twill pants, beige and red T-shirts, an Irish-knit sweater and beige deck shoes.

We had two small living areas where space was rather limited. One was our boat which we bought complete with green curtains, yellow tweed seating and brown carpeting. We redecorated our vessel before we left the mooring. The brown rug was exchanged for a bright navy one. The seating was reupholstered in a white, blue and red check, the curtains replaced with fresh white. I bought white dishes, blue towels and blue and white striped sheets for the bedrooms. My husband loves my nautical terminology. I used to describe our boat as having two bedrooms, a living room, kitchen and a back porch. The engine room was the basement. Anyway we both use **NAVY**, **WHITE** *and* **BLUE** *for our clothing, how can you tell? Our duffelbags are made of blue denim. Our clothes and stuff could go anywhere and they would blend nicely with the interiors. When we had some guests who fell into the* **BROWN** *category—well, that is another story.*

Our other small space was a 1913 Canadian Pacific Railway caboose which we restored and brought over to our island property. We used it as our island retreat until we built the main house and then it became our guest cottage. The caboose was thirty-six feet long and ten feet wide—just like our boat.

Our clothing consisted mostly of jeans for my husband, and denim pants for me—I owned some designer jeans but I prefer to be able to sit down sometimes. We wore T-shirts, sweaters, denim jackets, and windbreakers without stripes. **BROWN** *lifestyle,* **NAVY** *personalities and* **NAVY** *environment? Yes! We decorated our caboose with pale oak antiques, blue denim seating, and collected dinnerware and silver from CPR ships, trains and hotels. I guess you could call that "Surroundings in Harmony with Your Personality." When you have the opportunity to select a colour scheme for your environment to suit your personality, why would you not opt for it?*

Your Signature for your Garden

The entrance garden serves as an introduction to the people and the surroundings inside, while the back garden is a peaceful continuation in Colour and Style. Your balcony, patio, window box or garden, like your home, are a reflection of you. Just for fun look at the gardens in your neighborhood and try to figure out "who" lives in the house. I have designed many gardens and always considered the interiors of the home and the owners' attire when deciding on Colour, Texture and Style for outdoors.

Here are some fun guidelines for each colour personality. **NAVY** likes a neatly-trimmed lawn, clipped hedges and shrubs. A rose bed or two for the summer, tulips fill the bare spots in the spring, and pansies bring colour in winter. You will most likely find beautiful pots and urns overflowing with geraniums and other summer annuals at the doorstep or on the patio.

The **COOL-BROWN** collects unusual flowers in **PALE PINK**, **MAUVE** and **WHITE**. You will see grey shrubs of lavender mixed with other herbs crammed in every available space. Whether it is a balcony or a garden, the atmosphere is one of tranquility, enhanced by a small pond or a fountain and candles in hurricane lamps or lanterns.

The **BROWN** personalities are drawn to the outdoors by their own nature. Their gardens are full of flowering shrubs, perennial flowers, border annuals as edging in bright colours, and maybe a touch of **BLUE** for the **NAVY** person in the house. There is a always a vegetable garden.

The outgoing **BLACK** personality usually has a low-maintenance garden which is executed with precision. Clipped hedges outline the courtyard or garden, standard trimmed shrubs give it texture, and boxwood-bordered flower beds enhance the beauty of the roses. Big handsome pots, sometimes even in black, are often planted with ivy and tall pyramid cedars for the front door.

The **PALE BASIC** personalities think it is fun to draw from all the ideas and are sure enough of themselves to convey their own personality and style.

Gems from my Collection

Here are some true stories about gardens which convey the owner's personality. Samaya (**COOL-BROWN**) has an apartment which is decorated in **TAUPE** and **OYSTER**. Last winter she had a beautiful window box planted with white heather, dusty miller, white cabbages, white pansies with wine-red eyes and variegated ivy. Nestled into this profusion of blossoms were two little brass lanterns.

Alannah (**SILVER-GREY**) has a very beautiful home which is decorated in **WHITE** and **BLUE**. Her signature is **PINK**—all shades from pale to hot—but **NAVY** is her approach to how she conducts her life. "A place for everything and everything in its place." She chooses white clay pots and urns for showing off vibrant blue pansies in the winter, and pink petunias in the summer. These pots are showstopers and definitely suit her personality. She would not thank you for bringing her a red geranium or a red poinsettia.

My daughter Fiona (**TAUPE**) has a generous array of pots and window boxes on her balcony which are filled with pastel blossoms, hot pink flowers and herbs. In one corner is a bench covered with an Oriental rug and piled high with pillows. A little fountain adds to the atmosphere of this hideaway which is softly illuminated at night by lots of candles. It is a true extension of her apartment which is painted in **SOFT PINK** and **STONE** and is decorated with pillows, rugs, collectibles and of course, a generous display of candles.

I would like you to consider the colours and style of flowers you take or send to your friends' home. Mauve flowers for a **BROWN** household or hot pink gerberas for a **WARM RED** decor would not be suitable, neither are orange and yellow mums for a pale green and rose pink chintz living room. Don't you agree?

Our own garden mirrors our interior environment in Colour, Texture and Style. Our interior is French country—emphasis on "country." The Colours of the flowers complement the living room, dining room and kitchen. For Texture there are the trimmed boxwood shrubs in ornate clay pots, and slender pyramid cedars—I pretend that they are cypress—which add a bit of formality to the entrance and courtyard. One Florentine putti—which greets our guests at the front door—a fountain and a small stone bench nestle in the hydraengas in the back garden—set the style for our little sanctuary. I was pleased when I won an award for our garden at a recent competition.

The foregoing chapters about your personal signatures, in all areas of your life, will give you new vistas to communicate your personality and at the same time help you gain understanding of your family and friends. Each colour personality has its own style for expressing itself. When you have found your true colours which reflect a combination of several facets of your personality, you will also have your signature.

Start with the "bones"—the structure. Your living room has the seating arrangements as the "bones." The "bones" for the dining room is the table; for the bedroom it's the size of the bed; for the garden, the hard surfaces, like a patio; and last, the "bones" for a basic wardrobe is the colour.

The next part of my book, "The Blueprint," covers everything you ever wanted to know about a basic wardrobe suited for your lifestyle, and the appropriateness of the costumes for your stage. I will again address the different colour personalities and their signatures in all areas of their life.

As you read this book, you will come to realize that even your choice of a suitcase, the selection of your travel wardrobe or what you wear to a picnic is influenced by your personality traits which manifest themselves in everything you do in life.

One of my students phrased it so aptly: "Applying the teachings of the Blueprint in your life is the foundation of building an organized lifestyle. For doing that, a more pleasant and beautiful life will be your reward."

A SUMMARY

HERE IS A BRIEF SUMMARY OF WHAT YOU HAVE LEARNED IN the first part of this book and for your consideration:

- Realize that you are a person and not a people.
- Find one Basic Colour group to which you can relate most of the time.
- Experiment with texture in clothing and environment to convey a different image.
- Practise "letting-go" of an old image.
- Change your style of dress to suit your personality.
- Consider your environment.
- Continue your colours in your surroundings.
- Add accessories to your home which have your partners signature.
- Be open to new experiences.
- If your partner is **NAVY**, straighten the tube of toothpaste after you squeezed it in the middle.

It is the little things in life that count.

CB
A
CF
CF
fold

Introduction to the Blueprint

Welcome to the **NAVY** part of my book. The Blueprint contains complete guidelines for dressing appropriately for all occasions according to your personality, for all venues, in the right cut for your body, and having all your clothes in a closet designed for your personality.

You will love the how-to-shop section. I will wrap it up in a summary which should make you the best dressed, fun-to-be-with person around.

When planning your wardrobe suited to your personality, please consider your personality first. Identify the cut of clothing, not only the style, in which you feel most comfortable. Think about it. Do you like your clothes close to the body? Most of my **NAVY** clients do. Do you like a hugging neckline or do you prefer it to drop off your shoulder? Do you like a cuff at the end of your sleeve rather than a full, loose one? Do you prefer a blouse or shirt with natural shoulders rather than a dropped shoulder line? How about a narrow belt rather than a wide cinch one, and pants with a crease? I am speaking of your basic wardrobe and not the odd piece you may choose to purchase. Just don't let the new blouse you bought for that special occasion wear you—taking physical energy from you while you are supposed to be having fun.

Your Basic Colours are used to build a basic wardrobe and for the purchase of your "investment" clothing (furs, leather, cashmere, etc.). Buying shoes, stockings and bags in these Basic Colours makes life at home and travelling, simple and economical. Plan to have all your basic garments in these three Basic Colours in your wardrobe at all times.

The Parent Colour in your chosen colour group makes the strongest statement. The first impression you make in that colour is explicit. Only choose the best quality in fabric and construction for the garments in your Parent Colour.

The Number Two Colour (moving up the scale) gives you additional choices to communicate on different levels without losing the basic signature of your Parent Colour group. You are able to choose on what level you want to interact with people, and not be overpowering with your message. That is the reason that the number two Colour is the most important colour in your wardrobe. It is suitable for many occasions in your life—from active times to relaxed moments. It is the colour which takes you through the seasons and, as a bonus, it will match every item in your present closet.

Versatility and the ease with which it is integrated into your existing wardrobe is the reason that I choose the lighter Basic Colour to start your wardrobe.

The Third Colour is the lightest shade of your Parent Colour, which can be used as an accent to the darker Basics of course, but can also stand alone as a complete costume to convey a relaxed and open attitude toward life. (See the chapter on **PALE BASICS**.)

Your Accessory Colours are not limited to being used only in this way. It will be fun to apply your accessory colours—from light to dark—to portray a certain mood or a change of season. You may prefer to wear many different colours, using them as communication tools. An accessory colour can become your signature and then it makes an impact on its own. When wearing a whole outfit in one of the accessory colours, accessorizing is done with the Basic Colours. Let me use a red dress as an example. A **BEIGE** personality would wear beige shoes and accessories with the dress while a **NAVY** personality might wear silver grey shoes with her red dress, and the **BLACK** personality wears **BLACK**. Your choice of colour for your shoes makes the personality statement. Remember, always display your signature through Colour, Texture and Style.

The most important step is for you to decide on a colour scheme. Please select one of the Basic Colour groups which best communicates your personality traits most of the time. Stay within those suggested Basic Colours. Do not feel that you are going to be confined. Even if you are a person who likes all colours at some time or another, many options within each group are available to you.

The colour group you have chosen is based on your personality, lifestyle and surroundings, and can be used in varied shades from the darkest to the lightest. Choose your colours to communicate different moods and situations, from the precise (dark colour) to the relaxed (pale colour). Concentrate on your colour group to build a wardrobe suited to your personality.

Remember: you can make a first impression only once.

FOUR BASIC COLOUR GROUPS WHICH PORTRAY DIFFERENT PERSONALITY TRAITS

NAVY group	COOL-BROWN group
NAVY	COOL-BROWN
GREY	TAUPE
WHITE	OYSTER

BROWN group	BLACK group
BROWN	BLACK
CAMEL	SILVER
OFF-WHITE	WHITE

Wardrobe Planning for your Lifestyle

To own a wardrobe designed for your lifestyle and in harmony with your surroundings, does not take a big financial commitment; however, it does take good planning, guidance and discipline. If necessary, you can purchase just one piece at a time. I have always taught that it takes three years to build a colour-coordinated wardrobe. I will help you to have fun collecting clothes, and to have the correct garments in your closet in a very short time. Here are the three categories of clothing for you to consider. Remember to invest your money where your lifestyle is: City, Country and Recreation.

For the City/Business

Garments suitable for this category are related to the **NAVY** costume which is in general, more refined in fabric and cut. The City image includes all your business meetings, working at the office, bank or store or other activities within the business community. This costume also includes shopping downtown and having dinner at a fine restaurant. ("Lounging" under this category requires an upgrade from casual knits for special evenings at home.)

For the Country/Casual

The clothing for this category is related to the **BROWN** personality because of venues, texture in clothing and unconstructed style. Most people can relate to it. These clothes are meant for meetings at someone's home or yours, and for shopping trips in out-lying and rural areas. These clothes are also suited to wear to a family restaurant, having dinner with friends or going to a movie. What makes this image different from the City look is the fabric (Texture) and the Style which has movement, and does not cling to the body. (For an evening at home with family and friends, wear knitted pants with a matching tunic.)

For Recreation/Staples

Sports activities of any kind on land or water require the appropriate attire. This category addresses all leisure outdoor activities, as well as the environment of your cottage.

Use the same practical and washable clothes for cleaning and gardening. I call the clothes in this category "staples." You could buy them three at a time from a catalogue if you want to make shopping easier. T-shirts, jeans, jogging suits and runners are on your shopping list.

As you will have noticed, neither jeans or runners are listed in the City Category, and no outdated or worn out pants and sweaters appear in the Recreation column. Please familiarize yourself with these three categories of surroundings and the appropriate costumes for each of them.

These three outlines of different lifestyles explain the different settings and venues you might experience during the course of a day. The following chapter explains the different garment styles for each category.

SPLENDOR SINE

Basic Wardrobe Plan for Women

THIS IS THE START OF YOUR BASIC WARDROBE PLAN EXPLAINING each garment style and colour choice. (Find your colours first.) Don't pass over this chapter, it is a valuable lesson on how to build a wardrobe, using your present clothes, and building on them one by one.

The number two Basic Colour—**GREY**, **TAUPE**, **CAMEL** or lighter—is the best choice for a ***jacket*** when you start your new wardrobe plan. This jacket is the most important garment in your life; it is challenging to be without one. The colour, fit and style of your jacket are the main considerations. Your jacket has to be worn successfully over everything you own. Every colour of blouse and sweater you presently own must fit under the jacket and it must go with all skirts and pants of all lengths.

By adding only one of your new Basic Colours as a stabilizer, and without having to discard pieces of your wardrobe not in your new colour, you can start to build a new image. Suppose you have chosen **NAVY** as you Basic Colour scheme and your number two colour is **GREY**. Your grey jacket becomes the accessory to every coloured pant, skirt, blouse and sweater you own.

The fit and style of your jacket, which you will be wearing with several things and for many occasions, is of great importance. The shoulderline is the focal point, including the neckline or lapel. Look for good workmanship. More steps in construction increase the price of the garment, but will prove to be a good investment in the long run. Choose fabrics suited to your personality and check them for wearability throughout different seasons. The Style is as important as the Colour and Texture. Remember to select a jacket whose Colour, Texture and Style portray your personality, e.g., a grey flannel fitted or a **TAUPE** chenille wrap jacket.

Your new ***skirt*** might be part of a suit with your jacket mentioned above. Because we want to collect clothes like pieces of accessories, a fact I can't stress enough, a suit, per se, is too restrictive. You have to be able to wear the jacket over other garments. Always look at the individual pieces first (skirt and jacket). If they can stand on their own, then it is a worthwhile purchase. Skirt and jacket do not have to match, and please do not try too hard to match colour and fabric. A shade lighter or darker for one piece is often preferable. The skirt has to be in a versatile style as a "basic" to serve for different occasions. No! You will not wear out one piece faster then the other—they will both work overtime.

Slight fullness in a skirt makes it wearable with low heels and textured hose to suit both the Country/Casual Category and, with higher heels, the Business environment. The straight skirt is more or less limited to the City and Business activities. Choose your skirt in a fashionable length, short or long.

Basic style and fabric of ***pants*** suited to your personality in top quality material is your next purchase. The pants can be in your number two colour or select them in a slightly darker shade than the jacket. The **NAVY** personality might select a grey pair of pants with a sharp crease that fit fairly close to the body, touching the shoes. The **BROWN** personality prefers her camel pants in a more relaxed style with a pleated front and no crease. The person who prefers **BLACK** will most likely choose a slim-cut basic pant without any detail. The less detail on your pants (i.e., back pockets, cuffs, belt loops) the easier it is to wear them for different occasions—day and night.

Pants with back pockets fit into the Country/Casual or Recreation/Staples Category but are difficult to accessorize for City or evening wear. Pants with cuffs have that proper Business look but cannot be worn with high-heeled shoes, making them inappropriate for evening events. Belt loops on pants limit their use to specific belt widths, which might prevent you from using your signature belt. There is a lot to think about, isn't there?

I have just described a three-piece ***suit***—jacket, skirt and pant. You now have the base for a workable wardrobe. It becomes even more versatile when you choose all three pieces in a fabric suitable for the climate of your place of residence for most of the year. Light-weight gabardine or wool flannel have a pleasant weight and are right for travel, except in the tropics.

A ***sweater*** complements pants and skirts and should be part of a basic wardrobe plan. The sweater needs sleeves which fit under your jacket and a neckline which is versatile enough for wearing jewellery or scarves of different sizes. To add a cardigan sweater in the same or complementary texture really completes this picture. A silk shirt becomes dressy enough for informal evenings and, when worn with a silk skirt is suitable for visiting friends.

The crewneck in number one or number two colour of your Basic Colour scheme enables you to wear a different blouse underneath. A V-neck top could be in an accessory colour, e.g., **RED** or **PALE BLUE**—the neckline filled in with your favourite coloured scarf or jewellery. The last basic style sweater is the turtleneck or cowlneck. Both styles are prettiest in your number three or palest Accent Colour.

Make a beautiful ***blouse*** in number three of your colour scheme a part of your wardrobe. If you prefer a shirt, then may be you can find one with its own scarf or sash. A blouse transforms your basic suit, jacket, skirt or pants to express different moods more than any other garment. Silk or silk-like fabrics softens the strict **NAVY** or **BLACK** image and add a more refined look to the **BROWN** personality.

Consider buying a coatdress which can take the place of a suit and offers unlimited ways to accessorize. This garment has to be purchased in a size slightly larger than normal in order to have room for wearing a sweater or a blouse underneath. A two-piece ***dress*** in light wool is my choice for a basic wardrobe. You would gain an extra skirt to wear with your suit jacket, and the top may fit over your Basic-Colour skirt and pants—all from the basic capsule. You could wear this dress with a blouse or sweater giving you another outfit which could pass as a suit.

The colour for this two-piece dress could be either number one or number three Basic or one of the Accent Colours. If using Accent Colours, the colour must be versatile enough to be accessorized with others from your existing wardrobe. Executed in off-white wool or a silk-blend fabric takes the two-piece dress from an important interview or speaking engagement to cocktails and dinner simply by adding a camisole and leaving the top slightly open. The skirt of the same dress with a cashmere crewneck sweater, accessorized with jewellery, is a wonderful costume for a luncheon. The jacket over gabardine pants is an appropriate attire for a champagne breakfast in a hotel.

The two-piece silk dress is another favourite of mine. Try a number two Basic Colour silk print. If you can sew or know of somebody who does, do not stop at the blouse and skirt. Add a generous scarf or perhaps a large square shawl form the same fabric, as well as a camisole and full-cut pants for variety. These pieces will do overtime for you and you will not tire of them, I promise.

The basic ***shoes*** are all in number two or number one colour. You'll need one pair of high or medium-heel pumps, one pair of sandals for silks and other dresses, and one pair of low to medium dressy walking shoes suitable to wear with skirts and pants. Your handbags should coordinate with your shoe style, but please, do not match shoes and handbags in design. A bow on both is a bit like painting by numbers.

You will need a clutch or shoulder bag to go with your suits—neither of them oversized—in your Basic Colour number one or number two. A bigger carryall in leather or good-looking canvas in a colour to match your new two-piece wool dress is appropriate for many occasions, especially for the country look. This bag can hold your clutch or shoulder bag so that you do not need to take the carryall to a restaurant. One more bag is important for a well thought-out wardrobe: a small leather clutch just big enough to hold glasses, lipstick, compact, to go with your after-five wear, dinner at a restaurant, or a visit to a friend's home. When you dress for dinner, a jeweled evening bag is usually not appropriate with most garments we wear. The small clutch in number two or number three colour, i.e., **SILVER GREY**, **STONE**, **BEIGE**, or **BONE**, will do wonders for more occasions than you might think. I have a **RED** one, also.

When you look at my wardrobe plan, which has worked for more than thirty years, you will realize that you do not need many clothes for any given season. It is important to always have the basics (see chart) in your closet. You might have more in your closet when you start your planning but be sure that your basic wardrobe becomes your first priority to assemble. The wardrobe plan is designed to have about three pieces of clothing in each category for each season. By purchasing garments which do extra time—treating each piece of clothing like an accessory—you save on your wardrobe dollars. Try using your basic garments and making them into accessories. Let them move and not stay as a background. The **NAVY** personality has a challenge with this. Let go and try it; it works.

Wardrobe Plan For Women

GARMENTS	CITY/BUSINESS	COUNTRY/CASUAL	RECREATION/STAPLES
COATS	Fashionable	Raincoat*	Oilcloth slicker
JACKETS	Suit jacket Updated blazer	Cardigan* Tweed jacket	Nylon shell Windbreaker
SKIRTS	Straight, gored	Corduroy, tweed	Cotton twill
PANTS	Classic cut w/crease	Corduroy, leather Denim	Twill pant Jeans
SWEATERS	Crew, V-neck or cowl	Turtleneck* V-neck	Cotton knit Irish wool
BLOUSES	Silk or silk-like	Cotton shirts*	T-shirts, sweatshirts
DRESSES	2 pc. wool or silk/coatdress	Denim Cotton knit	
EVENING	Silk Pants	Wool or cotton knit	Jeans & cotton shirt
AT HOME	Tunic/blouses	Pants & top	Cotton knit pants & top
SHOES	Pumps in different heel heights Dress boots	Loafers* Sandals, boots Walking shoes	Running shoes Rubber boots
ACCESSORIES	Silk scarves Jewellery Gloves, leather Belts Briefcase	Cotton & silk scarves Textured jewellery Handstitched gloves Canvas or leather belts Shoulder bag & carryall	Cotton scarves Canvas belts Canvas or nylon duffelbag

*suitable for the city

Gems from my Collection

Here is a Gem to illustrate the act of collecting complete costumes versus making your basic wardrobe pieces serve as accessories.

Hilda came to my seminar many years ago. She is petite and beautiful with a clear and precise personality. Her colour scheme is **NAVY**, **GREY**, **WHITE** and using bright Accent Colours. I always saw her in **SILVER GREY** and **PASTELS**. Not long ago I had a private consultation with Hilda in her home. She lives in a beautiful house which is decorated mostly in **WHITE** and **PALE BLUE**. Nothing is out of place, everything is colour-coordinated, and her closets are in perfect order. This is truly a **NAVY** personality who enjoys order: "A place for everything and everything in its place." Interestingly though, she had bought everything in sets, which, of course, is a **NAVY** trait. She owns jackets with trim that matches the pants and the T-shirt has the same trim as the jacket. Hilda even arranges each costume the way she will wear it. Blouse, skirt, belt, scarf, jackets, pants, shirt, belt and handbag—no mix and match here. No, she did not use her basic jacket as an accessory; it stayed right where it belonged with the matching pants or skirt. We laughed about her **NAVY** traits and started mixing and matching her costumes, and she became excited to see the many different looks we created come to life from her basic wardrobe.

I noted something else: her clothes were mostly **GREY** and **PASTELS** and her husband's clothes were **BLACK** and **RED**. That is probably why there were wine-red towels in the bathroom and black ones in the powder room. I asked her about it, thinking that her husband liked her as a pastel lady. Not so! He wanted her to wear flashier clothes and gold jewellery. She prefers dainty silver pieces. Hilda is not a timid woman at all, she just kept buying grey and pastel pieces of clothing. I told her that she did not need to do that to herself. She could show her forthright manner with some colour. I suggested a bright blue coat for her silver-grey wardrobe, and on her recent trip to Greece, her husband bought her a raspberry red leather jacket. What a great accessory!

Now back to the basics. I suggested a jacket, a skirt, a pair of pants, a sweater, a cardigan, a blouse and a dress. Add some shoes and jewellery and it would all fit in one suitcase. That is basic. You and I would most likely not need much more than that, I guess. I read a book recently on how to minimize one's lifestyle, by making one's wardrobe: basic, make-up: none, surroundings: barely, and some other lean suggestions. By the time I got to page 85, I was so depressed that I needed an antidote, and sat down with my *Victoria* magazine. (No! Not *Victoria's Secret*, even though that could have been alright, too.) I believe in a basic colour-coordinated wardrobe but I am not a minimalist. My wardrobe plan consists of fewer pieces in your closet rather than a closet full of clothes and nothing to wear. However, it is fun to collect items which work best for you when you mix and match them which gives your wardrobe your personal signature. It's all in the planning.

Buying clothes without a well thought-out, coordinated wardrobe plan is like building a house without a blueprint.

Basic Wardrobe Plan for Men

Although the majority of my students and clients were women (about 2000 every year), many male clients came to my special seminars. I also gave private consultations in their office or mine, and worked with some of the staff of large companies. My questionnaire for men was the same as the one I used for women, and I found it interesting how detailed the men's answers were to the questions. Most men were interested in their personal environment and even more so in the image they portrayed in their business and private lives. I remember one of my breakfast meetings where several men lingered well after my talk was over. They wanted detailed information: "Should I have different frames for my glasses? Shall I grow a mustache? Do you think these shoes are appropriate for the office?" And on and on. I truly enjoyed being with them—and they took my theory of colours and what they represent seriously.

I sincerely hope that you will enjoy learning about the image you portray or want to adopt. All the approaches to dressing in harmony with one's personality applies to men as well as it does to women.

The following chart is a guideline for a basic wardrobe plan. Choose one of the colour charts and the related personality traits with which you can identify most of the time, and plan your wardrobe to suit your personality in Colour, Texture and Style.

The ***City/Business Category*** on the following Wardrobe Plan for Men descibes the **NAVY** personality, however, you will choose your colours. Most of the dark Basic Colours are suitable for business: **NAVY**, **GREY**, **BROWN**, **TAUPE**, **BLACK** and even **DARK GREEN** (a colour which suits a **BROWN** personality's business attire). Depending on your job, you need three to five suits and one blazer with two pairs of pants. The blazer is an important garment for most men because it is easy to wear and appropriate for many occasions. The blazer fits most lifestyles and surroundings—from city, to country, to travel.

The ***Country/Casual Category*** of garments has the look to which most people can relate. This look conveys a more relaxed City image not to be confused with the next category designed for hiking and camping activities. These country clothes should convey the "I-care-how-I-look" message, one expects in the City/Business attire. This Country/Casual arena is where you can make your signature of always being well-groomed and appropriately dressed. Don't ignore this!

The ***Rectreation/Staples*** in your wardrobe are exactly that. You could buy them in threes. I do not have to explain what kind of clothes you need, you are familiar with them, but look at the outline to complete your list. One thing to consider: if you want to be known as "well-dressed," please wear plain T-shirts (without a large logo or picture) and plain runners for "ordinary" activities (without coloured stripes). Thank you!

There are three different categories of jeans, each of which is appropriate for different environments.

#1 Jeans, regular western cut, are designed for cowboys, and are meant to be worn for outdoor activities: washing your car, hiking, cutting wood, etc. No problem, as long as they don't have holes in them. #1 jeans are for Recreation, but don't make it for the City look.

#2 Jeans have been pressed or they may be designer jeans which are more refined in fit and style. These jeans are appropriate for the Country look if you wear them with a casual shirt and loafers, not with athletic shoes. This outfit can easily be up-graded by adding a blazer. #2 jeans can be worn in the city as long as they are accessorized with the appropriate shirt and/or sweater, but are not worn for business. Can you help to move casual Friday to Saturday?

#3 Denim Pants, are regular pants made from denim fabric and are appropriate for the Country/Casual Category.

None of these three categories of jeans is ever appropriate to wear in elegant restaurants or the dining room at a private club and, in fact, they are often banned. I for one endorse this policy 100 percent. Would you wear jeans to a formal dinner at your best friends home? I think not.

I use basic wardrobe pieces to create different images. Let me explain the looks from structured (**NAVY** personality) to unstructured (**BROWN**) using a blazer.

- Gabardine slacks, a sport shirt, blazer and loafers (**NAVY**).
- Cotton twill pants, a silk sweater, blazer, and city sandals (**TAUPE**).
- #2 jeans, knit sport shirt, blazer and clean runners (**BROWN**).

A crewneck sweater can replace the blazer for a more casual look. Shirts and shoes make all the difference. Shoes are really the tell-tale of the personality and style of the wearer. I am a shoe watcher. The following examples are based on observations of what a pair of shoes can tell.

Three pastors dressed in their robes were each standing at their pulpit. The first, a pastor of a very large church wore a blue robe and visible below the robe were a pair of grey striped slacks and black shiny Oxfords. When I saw him later mingling with people over coffee, he wore a grey striped business suit, white shirt and tie. No surprise to me, he is a **NAVY** personality. The second, our own pastor in a smaller church wore black slacks and black loafers under a white robe the first day that we visited our new church. My guess was that he would be a **TAUPE** personality. A few days later he and his wife came for a visit to our home. He wore black slacks, black loafers with a single gold bar across it, a taupe silk shirt—no tie—and a taupe and black tweed jacket. A perfect outfit for a **TAUPE** personality. On our visit to Arizona, we went to a small church in the mountains of Sedona were we observed the third example. The priest wore a brown robe and sandals. **BROWN** was easy to guess. When he removed his robe, he was in harmony with his surroundings. I wanted to end it here but my **NAVY** husband thought I should mention that the priest was dressed in jeans and a T-shirt with his brown sandals. It is true that your attire will get its signature from the shoes you wear. Running shoes say "casual". How "casual" do you want to be?

Speaking of dressing casual, do you rise to the occasion and wear a suit when your partner is dressing for an evening out? Do you wear a tie or are you following Prince Claus of the Netherlands and wear your suit without one? I agree with someone who said, that a suit without a tie is like a car without hubcaps. Of course there is always the one exception like when a handsome man wears an Italian-style taupe suit, a silk shirt with a banded collar and elegant sandals. No socks! Oh, well!

Wardrobe Plan For Men

GARMENTS	CITY/BUSINESS	COUNTRY/CASUAL	RECREATION/STAPLES
COATS	Seasonal coat	Raincoat*	Oilcloth slicker
SUITS	Worsted wool Pinstripe/glencheck Flannel	Corduroy Tweed	Jogging suit
JACKETS	Jacket Updated blazer	Tweed* Cardigan,leather	Windbreaker
SLACKS	Plain or pleated Cuff or no cuff	Corduroy, denim Cotton twill	Cotton twill Denim
DRESS SHIRTS	Plain/striped	Button-down pinpoint*	
SPORT SHIRTS		Plaid, knit	T-shirts
SWEATERS	Crewneck V-neck/vest Silk sweater	Turtleneck V-neck, or crew Sweaters	Sweatshirts All sweaters
SHOES	Oxford/tie Slip-ons	Loafers* Low boots	Athletic shoes Rubber boots
AT HOME	Wool twill pants No collar shirt Sweater	Pressed jeans Sportshirt Sweater	Jeans & T-shirt (no logo)
ACCESSORIES	Silk ties & hankie Leather belt Leather briefcase Cufflinks	Wool tie*/assorted Braided belt Leather or Canvas carryall	Cotton scarves Canvas belt Nylon duffelbag

*suitable for the city

Guidelines for Women over Sixty

This chapter is for women who think they will be old ten years from now but who would like to feel and look their best as long as they live. Middle age was defined as being thirty-five years old not too long ago. Today, we like to think that middle age begins around the age of sixty. We are more health-conscious, look younger and work at staying healthy and in good shape.

"*Just one more time*" enroll in an exercise program, join a health spa, swim or go for brisk walks. Nothing will age you more than poor posture and a shuffle disguised as a walk. Swing your arms, let the hips and arms move as one. Portray youth by the way you sit, stand and walk. It has more sex appeal.

"*Just one more time*" change your eating habits for the better. Lose extra pounds and you will feel more energetic. Just think: if you gain one pound after Christmas every year starting at age thirty you would be thirty pounds heavier by the time you're sixty. Or do you believe you will be old ten years from now, and then it would not matter? I have news for you, it does matter. It always does.

"*Just one more time*" give your mind and body a wake-up call.

"*Just one more time*" have a facial. Don't give up on skin care ever. Indulge (a North American term) in a monthly facial done by a professional or learn to do it at home. In Europe a facial is simply an essential part of an all-over grooming program. Every inch of you—your face, your body, hair and nails has to look well-groomed to make a youthful appearance. This is important at any age.

"*Just one more time*" become aware of make-up trends. The mature face needs a fashionable and sophisticated make-up. Do not create the illusion of age by not wearing make-up at all or by wearing an out-of-date face.

"*Just one more time*" ask a good hairdresser to create a youthful cut and a style suitable for your kind of hair. A modern hairstyle to fit your young personality is your aim.

"Just one more time" purchase a well-fitting bra for an uplifting experience. It is of great importance to wear shoulder pads with your garments (1 cm/¼" for blouses, 2cm/¾" for dresses and jackets). Shoulder pads camouflage rounded shoulders and make garments drape over your body rather than cling and fold in the wrong places. Most of all, it will make you look youthful—straight shoulders always do.

"Just one more time" look at fashion. Do not give up on smart, fashionable, quality clothing by being a bystander in your costume from yesterday. Do wear clothes with the appearance of youth. Action is related to youth. In order to look younger wear sporty-looking clothes that provide movement and style. Seldom do runners look as though you are training for a marathon, but rather that you have sore feet.

Most of all think colour. Action colours. **RED**, **BRIGHT BLUE**, **GREEN** for the **NAVY** personality. **HOT PINK** for **GREY**, **PERIWINKLE BLUE** for **BEIGE**, **PURPLE** for **STONE** and all **BLACK** personalities, and **TOMATO RED** for **OFF-WHITE** or for **BROWN**. Start collecting bright-coloured garments to add to your basic wardrobe: colourful shirts and sweaters, and a smart-looking coat in bright **BLUE** or **RED**. *Always think colour.*

Watch for colour trends in the market place, but do not succumb to using them "just to be in," unless they are "right for your personality." You may wish to add just a touch of the new **LIME-GREEN** or **HOT PINK** in an accessory.

"Just one more time" get interested in accessories. Purchase a coloured umbrella or bright gloves as the only colourful touch for your basic coat. Wear pieces of jewellery in a more unconstructed way: plain ivory (**PALE BASIC** personality), jade (**COOL-BROWN**), or black coral rings interchanged with your diamond rings to pick up the colour of a garment. Hang one of your gold charms (from the charm bracelet of yesterday) on a silk cord to wear with a T-shirt or sweater. Wear your diamond stud earrings with jeans. Mix Venetian beads with your pearls, and colourful plastic bracelets with your gold or silver ones. For Christmas attach a small angel on your gold necklace, and wear a lapel pin anywhere—on your shoulder or on your cuff—but not on a lapel. Think young, have fun!

Gems from my Collection

Here are some examples of women in their sixties and how colours played a role in the way they portrayed a more youthful image. These women became more energetic-looking because of Colour and Style changes, and even felt more at ease with themselves.

Mary is a striking silver-haired women in her sixties who came to me for a consultation. She is outgoing, travels a lot, and is interested in people and events. However, she did not feel wonderful in all the pastel colours she was wearing. **MAUVE** and **PALE BLUE** are the accepted colours for the silver-haired, sixty-plus woman. Clothing sales persons will readily agree. But what happens if you are not a **MAUVE** personality? Part of your daily energy goes towards overcoming the low-key image and convincing people you are still alive, are in the stream of things and have no intention of retiring. Mary is still having lots of fun planning and living.

We agreed on keeping **GREY** as her Basic Colour but used it more as a canvas on which to play all her new bright colours. **RED** became the predominant Accent Colour in her wardrobe because of her temperament and zest for life. When I purchased a bright raspberry-red raincoat for her, it put a spring in her step when she left my office.

Helen, a shy, little lady also in her sixties, was a different story. She had raised a family and her husband was still actively involved in business. He had decided to re-carpet the whole house in **RED**! Helen was quite upset about his choice of colour. To add to all this fire and excitement, she wore a crisp and close-to-the body black suit with a starched white blouse. The outfit was a present from her husband. I don't know how she had the courage to come to me for a consultation.

We had a wonderful time talking about her feelings regarding colours and how they affected her. My recommendation included not only to use colours and clothing suited to her personality, but also make some changes to her environment.

She is now wearing softly textured clothes in **SILVER** and **GREY**, as well as some lovely pastels so well-suited to her personality. To overcome the red carpet in her living room and to create a more becoming background for her leisure hours, we decided to arrange large pastel pink, pale blue, silver grey and soft yellow silk cushions on her couch. I suggested turning down the lights, use lots of candles and to enhance the decor with plants and pastel flowers. We eliminated her tailored clothes from her wardrobe, and Helen, in harmony with herself, can now relax in her soft colours until she is ready to wear bright ones again.

I guess you have heard it said that **PALE PINK** *is only for the very young or the very old. Psychologically speaking that is correct because* **PALE PINK** *connotes frailty. I broke an ankle several years ago. (I tell people it happened while roller blading.) Beside the pain, I also felt that I lost my identity as an action-oriented person. It was a challenge being on crutches and later I had to walk with a cane. During that time, something interesting occurred. I was drawn to the colour* **PINK**. *I was still on crutches when I made myself a pale pink suit and in the spring I planted pale pink and white geraniums. Unheard of for me. I am the original bright red geranium girl. (Who would like to have a pale pink suit?)*

Guidelines for Children

INTERIORS OF NURSERY SCHOOLS AND KINDERGARTENS ARE USUALLY painted in bright, happy colours because children with their boundless energy are attracted to them and like to wear them. Why dress them like adults in **BLACK** and **PURPLE**? I even suggest that mothers of young children, kindergarten and grade school teachers also wear bright colours or accessories. Of course, not all children are outgoing and athletic. Just because their hair colour is blond or brown is no reason to dress them in the generally accepted colours, i.e., **BLUE** for blondes, **RED** for brunettes and **GREEN** for redheads. Several of my friends and clients were so excited about my colour theory that they included their children in our consultation sessions.

To dress children in colours suited to their personalities is as valuable to them as it is for adults. The following information on colour can be applied to children age five and up.

NAVY: An outgoing, active youngster interested in sports—girl or boy—looks great in the primary colours with **BLUE** undertones. They love jeans and bright T-shirts, especially striped ones.

COOL-BROWN: A shy girl who likes to be on her own is drawn to **PASTELS** and pretty blouses and dresses. A boy likes a white or pale blue T-shirt rather than a bright striped baseball shirt to wear with his cotton pants.

BROWN: The young person who prefers to dig in the garden, plays with flowers or snakes, loves the outdoors and going fishing with dad, is drawn to the colours with warm undertones; **GREEN**, **YELLOW**, **RED**. Give them corduroy pants or jumpers and tops that are soft to the touch; they will love it.

BLACK: The enthusiastic little **BLACK** personalities like everything and anything in bright colours. One can see and hear them coming. Have some extra T-shirts, bathing suits and socks on hand because they might "misplace" some items in the excitement of it all.

The Closet Space

Knowing children's personality traits presents the opportunity to create a closet space suited to them which they will then more likely use. It is easier if they have their closet space organized with boxes, baskets and shelves. If children like to display all their little treasures, they will also enjoy displaying their clothes on shelves (**COOL-BROWN**). Others make their room neat by stuffing everything into drawers (**BROWN**). Good thing they like jeans and sweatshirts. Bright plastic boxes placed on the closet floor make good organizers for the active youngster (**BLACK**) who would throw things into them on the run. The **NAVY** children might have a closet with hanging space and shelves behind doors so the room will (mostly) look neat. The sooner children are introduced to the fun part of organizing their room, the more natural it becomes to keep order.

Wardrobe Plan

Students and clients alike have consulted me about their children's colours in order to make shopping easier. The wardrobe plan for children is based on the same principles as that for adults. Choose three each of the main garments in each category, i.e., *School:* dresses, pants, shirts and blouses; *Play:* jeans, T-shirts and sweatshirts. Do not forget special occasions such as theatre, dinner and church. Consideration should be given to colour and its effect on the child. An outgoing child feels good in bright colours and a shy one does not. Bright colours take more energy to wear and stimulate the child. I would not put an active child in red pajamas. They could be awake longer than you like.

Imagine a shy, dark-haired little girl arriving at a birthday party in a bright red dress. When she enters the room, all eyes are on her, and she probably would like to sink through the floor. A soft pink or yellow dress would have been more appropriate for her personality, and certainly would have made her feel more at ease and able to be herself. Consider the blonde five-year-old tomboy in a pale blue dress with ruffles. Would she not prefer a red cotton jumper and white blouse?

The next chapter is devoted to the subject of travel—involve the children in how to plan their travel wardrobe and pack a suitcase. Have fun planning your next trip.

Travel Wardrobe and Luggage

Your travel wardrobe should emerge from your closet. Review the chapter "Wardrobe Plan for Your Lifestyle" and the descriptions of the clothes. If you feel the need to purchase new things for your trip, only buy clothes which are going to be part of your wardrobe plan—not clothes for the trip only. Nothing is more useless than spending money on wash-and-wear items when you prefer to wear other fabrics in your every day life (except when you go rafting in the Grand Canyon). Let the travel wardrobe plan chart be your guide.

Place all the clothes that you think you want to take on your bed. You will only need two of each type of garment for a four week trip. Believe me! Don't forget to consider the climate and unexpected temperature changes. The chart has a "warm" and "cold" column. My plan recommends taking two extra pairs of pants and tops for that reason. When you fill out your chart, place a mark for each one of your chosen garments in the appropriate column. If a garment fills only one space on the chart leave it at home. Don't take a dress which you can only wear at dinner—instead accessorize a blouse.

How well you pack a suitcase depends on how well your closet is organized and coordinated. Pack only two colours when traveling—more makes life too complicated. Think of each garment as an accessory. Remember to think of layers. A blouse could double as a beach coverup or as an evening jacket over a matching camisole. A coatdress could be used as a coat over a two-piece silk dress or worn unbuttoned to the waist over a T-shirt during the day. Your raincoat will get you "down the hall" in a small inn when you are wearing your pajamas (or, perhaps, nothing).

It might be of help, and it is fun, when you create dummies of your different looks right on your bed before you pack them. Check how the many ways your clothes will work. Add some scarves and jewellery. You will prove to yourself that the plan is functional. Do not be tempted to pack more clothes. I have had mail from all over the world thanking me for suggesting this practical plan.

When you pack for a big/small city trip, use a square suitcase. Everything is packed flat. If everything is flat and folded, opening the case is like opening a drawer: items are easily lifted out without disturbing everything else. Your shoes should be slipped into special bags in your colour scheme. (My shoe bags are **RED**, my husband's **BEIGE**.)

Accessories are important for your travel, they can change your daytime outfit into one appropriate for evening. They can add a touch of drama to what you are wearing. A shawl not only gives you a bit of panache in the evening, but it also adds warmth during the day when you cross the Alps or are on the plane. Cotton scarves are versatile. They can be used as a belt, worn different ways around your neck, or make neat little tablecloths when you stop for a picnic along the roadside, so take more than one.

Travel Wardrobe plan

	GARMENT	BIG CITY	SMALL CITY	RESORT	DINNER OUT	WARM TEMP.	COLD TEMP.
COVERING	Shawl	✓	✓	✓	✓	✓	✓
	Raincoat	✓	✓	✓	✓		✓
JACKETS	Tailored Blazer	✓	✓	✓	✓		✓
SKIRTS	Straight	✓	✓		✓		
	A-Line twill		✓	✓		✓	✓
PANTS	Garbadine	✓	✓	✓	✓		✓
	Stretch	✓	✓	✓			✓
	Twill	✓	✓	✓		✓	✓
	Twill (additional)	✓	✓	✓		✓	
BLOUSES/ TOPS	Silk	✓		✓	✓		
	Silk knit tunic	✓	✓	✓	✓		✓
	T-shirt	✓	✓	✓		✓	✓
	T-shirt (additional)	✓	✓	✓		✓	✓
SWEATERS	Crew	✓	✓	✓	✓	✓	✓
	Cardigan	✓	✓	✓	✓		✓
DRESSES	2 pc. silk	✓	✓		✓	✓	
	Denim coat dress		✓	✓			✓
LOUNGE/ EVENING	2-pc. knit		✓	✓			✓
SHOES	Low pump	✓	✓	✓	✓	✓	✓
	City walking	✓	✓	✓		✓	✓
	Country walking	✓	✓	✓		✓	✓

Your Luggage

Choose your luggage as you would choose your clothing to suit your personality and lifestyle, and to be in harmony with your surroundings wherever you are. A square suitcase, softsided or not, is meant for the **NAVY** personality who prefers structured clothes—gabardine skirts and slacks, shirts and blouses which can be neatly folded and which arrive at the destination with the least amount of wrinkles. This square suitcase also acts as a drawer for those who like their clothes to be readily accessible when travelling by car.

The zippered canvas duffelbag (in one of your Basic Colours, please) is favoured by the **BROWN** personality who prefers T-shirts, sweaters, twill or corduroy pants and wears loafers. This duffelbag is suitable for anyone who spends a weekend in the country or travels by car—inn-hopping perhaps. The nylon sport duffel is meant for the Recreation/Staples Category holding jeans, T-shirts, runners, etc. This is the bag you bring onto a sailboat or take to the cottage visiting friends. It is not meant for a stay at an elegant hotel or private club unless of course it is the extra bag which contains your jogging gear. (I would not have it in purple either.)

A zippered garment bag is handy for car travel to hold your blouses or shirts, skirts or pants. Purchase one in the same colour as your duffelbag if possible. Your car interior will look less confusing.

A full lenght wardrobe garment bag is useful for the business traveler. It holds an extra suit, shirt and blouse. You can pack the rest of your belongings in the outside compartments. The only consideration is the height—your height—you have to be taller than 5'3" to comfortably carry a bag like that.

My husband and I travel a lot by car and plane. We found the suitcases which suit our lifestyle and our personality. We both like our clothes neatly folded and accessible without having to look for our underwear stuffed in our shoes. We both choose square, softsided, beige canvas suitcases with our initials. The size: 20" x 15" x 9". Our duffelbags match. This is the luggage we use for a month-long trip. My husband's smaller duffelbag serves as carry-on luggage and I use a red leather bag light enough for me to carry when it is full. That's it! No plastic bags, no garment bags. For our two-months travel through Europe we both take a softsided, square, beige suitcase with wheels 26" x 18" x 9". We also take along an empty plain nylon bag to bring home our purchases.

Gems from my Collection

Wherever I am, every day of the week, I collect ideas and observe people and their clothes. I am a people watcher—which has helped me to write this book. I also make notes about clothes that work for me and my husband, and those that don't, especially on our travels. Perhaps the following little Gems will explain.

This happened about twenty years ago when we visited Switzerland and stayed at a country inn high up in the mountains. Each day we went for our morning constitutional: an hour walk-up. We wore shorts and running shoes. We were doing our aerobics strutting up the mountain path when we met up with a group of sedate (you don't "strut" up the mountain) middle-aged Swiss hikers all properly attired for the outing. Just as we passed them on the trail, we heard somebody calling in German: "Tennis Anyone?" Didn't I tell you to be in harmony with your surroundings?

The other incident happened in Salzburg. After spending a few weeks in Europe where I wore my beige and navy more or less tailored clothes with my gold L.A. Gears (great for city walking), we arrived in Salzburg on a Sunday. In a big city on Sundays one rarely sees tourists but certainly encounters locals. My gold running shoes were ever so wrong. I wrote in my notebook: Don't wear gold shoes in Salzburg! Salzburg is a "suit" city—suits worn with appropriate accessories. The next time I wore my tan-coloured highly polished loafers with a classic beige pantsuit and a large scarf with hearts printed all over. It did not take much to fit in did it? Anyway we do not travel with white runners unless we stay in a beach resort where I can wear them with twill pants, an off-white T-shirt and a cashmere cardigan over my shoulders.

We traveled to New York, Boston, Cape Cod and Pennsylvania not too long ago, and arrived in New York in late August. The weather was pleasant and not hot at all. The city was full of tourists wearing running shoes, T-shirts and jeans. When we returned to the Big Apple six weeks later, there was a different atmosphere on the streets of Manhattan. It took me a while to realize that all the running shoes and jeans were gone. The tourists had left! New York was New York, but we were experiencing it on a different level.

Even though I have not mentioned what I wore in New York, I certainly did not wear jeans and running shoes. However, I could have worn my gold ones. Look at my travel wardrobe plan and visualize the different pieces and then fill in the items from your wardrobe. I have not mentioned current fashion items because the book would become dated too easily. When I say: updated, that means current style. When you wear **NAVY**, you know that costume represents the city. Use Accent Colours in bright shades for the tropics and earth-tones for the desert. Enjoy your surroundings and become part of them. It takes so little to show an "I-care" attitude.

When you read the chapter on appropriate costumes for different occasions and venues extend those guidelines to travel. Let us agree to dress in harmony with your surroundings so when in Rome, do as the Romans do. Buy a bottle of wine? Yes, that also but dress in Rome or other cities to blend in. Jeans and runners will not do. Police suggest that for your security, do not obviously look like a tourist. Number one: wear comfortable walking shoes in stead of runners.

*Another gem from my travels occurred one time after conducting seminars in the city, I was on my way to our home on Galiano Island where we lived for fifteen years. (For my east coast U.S. readers, Galiano is **not** Nantucket!) I was stopped by one of the ferry officers during the loading procedure: "Lady, lady, this ferry goes to the Gulf Islands only, **not** to Victoria!" There I stood in all my **OFF-WHITE** splendor, and compared to the other passengers in their outfits, I looked rather out of place.*

Colourful Picnic Experiences

You do not have to travel far to enjoy the experience of visiting a different place. My husband and I love picnics. A picnic in the park, on the beach, or at other unique and unusual places may be the start of making your travel plans to destinations further afield. I sincerely believe that by breaking the routine of daily life with a small event like a picnic, you touch base with your partner—away from his office and your kitchen sink. Whether you plan to go by car, bicycle or by rowboat does not matter, but taking the time for the little pleasures in life matters a great deal. Plan a picnic with a theme that suits your personality.

Navy

An ***early morning picnic breakfast*** held on a viewpoint in your own area can be fun for those who like to get up early and watch the sun rise. A bright canvas bag can hold a checkered tablecloth and napkins, mugs and plates in red, blue or yellow with a matching coffee thermos. You are dressed in blue denim pants, a turtleneck sweater and colourful windbreaker. Your companion rose to the occasion as well and wore bright colours to meet the new day.

We have done that many, many times (even when we had full time jobs). We love sitting on a log on the beach, looking for shells or going for a brisk walk when the rest of the world is still asleep. What a great way to start the day. Even though we have a beautiful garden we enjoy being with each other in the stillness of the morning looking at the ocean.

Cool-Brown (Taupe)

If you decided that the **COOL-BROWN, TAUPE** and **STONE** personality traits relate to you, then you will probably imagine different settings for your special picnic. A ***romantic picnic*** in a rose garden in a nearby park might be your choice. You could pack an antique basket with delicate china, your ornate silver flatware and an embroidered linen tablecloth. You can sit on the grass on an antique shawl and watch the sun go down listening to tapes of Corelli. You may wear the faded-rose print cotton skirt—the one you found in an antique shop—with a cotton jacket. I did not make this up. Someone I know created this setting for the too-busy-for-anything husband; it was their first picnic of many.

My friends gave me a birthday picnic in a rose garden in our city, and we all wore hats.

Black–Off-White

A ***sophisticated picnic*** with crystal wine glasses, your good china and a linen tablecloth on a park table under a tree might be to your liking. Michella even took her silver candelabra along. If you are planning an event like this, you could wear the tuxedo shirt you found in a men's wear sale to top your new black designer jeans. The picnic basket, especially purchased for this occasion, is outfitted with compartments to hold cutlery and containers. You will use it frequently and perhaps, for your next picnic on top of a mountain, carried there by helicopter. I know a person who flew his girlfriend up the mountain for a catered champagne lunch to propose marriage to her.

Have fun dreaming up ideas for a theme which conveys your signature or particular mood. Make a picnic—like anything else you do—an event. Picnics certainly became such an important event for a group of people I know, that they formed a Picnic Club. Formal picnics in unusual locations even took place on the grassy median on the way to the Vancouver B.C. airport, and one member of the picnic club rented a Greek villa for another picnic. She spent her inheritance to fly her friends to Greece for the occasion. Now, that's a picnic!

Picnics have always been part of our life, and are the small interludes which add enjoyment to our daily routine. I want to share a special one with you. On one of our trips to Venice many years ago we had to make a choice between going out to dinner or to go with a gondola on the Grand Canal. We could not afford both. We decided on the gondola trip. We bought bread, cheese, prosciutto and a bottle of wine, borrowed large white napkins from the hotel and the gondolier supplied the red roses and the songs. We dressed for dinner. Needless to say it was an unforgettable picnic.

So was this one. My husband and I went on a car trip for a weekend in the country. We prepared potatoes, fresh vegetables and sausages, wrapped it all in double aluminum foil and my husband placed it on the manifold of his Corvette. When we stopped half way and opened the hood of the car by the side of the road, a motorist stopped and wanted to know whether he could be of assistance. My husband told the perplexed fellow traveler that he was just turning over the potatoes. In four hours we had a beautiful hot meal, overlooking the Rockies. (For cooking fresh pork you might have to travel another hour.)

What does this information have to do with the psychology of colour? Everything. Remember the small things in life make the difference; they give your life Colour, Texture and Style.

Appropriate Costumes for your Stage

Dress for yourself first. To be well dressed is a matter of pride and has nothing to do with finances or fashion. Be pleased with the way you look at every hour of your day. It is a priority that you feel good about yourself, how you dress and thereby set an example to the ones close to you.

Dress for your family and friends. To make an extra effort to dress for the people closest to you just makes sense since it is based on your consideration for others rather than hiding behind the comfort of: "It is just family and close friends."

Dress for the world. When you take personal pride in your appearance (always!) you dress appropriately, for wherever you are, and consider your family and friends in the way you dress. Then it becomes natural for you to show respect for all situations which come along in life.

This chapter deals with what to wear for situations arising from six o'clock in the morning to midnight. Each "act" needs an appropriate costume for practical and aesthetic reasons. When I first taught this lesson there was a lot of groaning from my students who saw themselves changing clothes six or more times during the course of the day. The reason for mentioning the following guidelines is so that you may learn what certain clothes communicate to others, and how being in harmony with your surroundings and lifestyle adds to the enjoyment of life.

Act I—Intimate Apparel

The first garment one reaches for in the day is usually the housecoat. Some women get into their jogging suits as soon as they get up, but if you were to purchase a housecoat, let me suggest the right one for you.

A colourful washable housecoat with a front button or zipper closing, and three quarter-length sleeve is suitable for the woman who has a husband, small children and an action-filled kitchen. The colour should be cheerful, and the styling one that does not catch on drawer knobs or sink faucets, and manages to stay out of the porridge.

A cotton or silk robe in classic styling to match your pajamas is another good choice for an active person and this ensemble is perfect for travelling, especially when you stay with friends or family. A low-heeled leather mule is the appropriate shoe.

A terry bathrobe is a wonderful garment for the time between shower and getting dressed or for relaxing after a bath or sauna. However, it is a bit cumbersome for doing things in the kitchen because you may have to adjust the wrap and tie belt just when you are in the middle of pouring your cafè-au-lait. The white terry robe could be a staple in anyone's wardrobe—it is especially nice to have an extra one or two for the guest room as well. It is, however, too bulky for travel. Terry cloth slippers are a good choice to go with this robe.

If you are married and have children, remember that the impression they have of you at breakfast will go with them when they leave for the day. Children are very much aware of what mother is wearing. Your husband might compare you with the women he sees on his way to work or in the office "all dressed for business," looking quite attractive. Do *not* let his last glimpse of you in the morning be of you wearing a long nighty, short housecoat, and wearing a pair of his socks. I did not make this up either.

To own a feminine robe in pastel coloured or off-white silk or silk-like fabric with matching gown is the dream of most women. A pair of silk slippers with a low heel complete this picture for special occasions. Please dream up a lot of them for your "stage." Whether you are alone or with someone special, or taking a trip on the Orient Express, it is a perfect costume for intimate surroundings.

The long T-shirt is the correct attire for the person who does not use a housecoat, but likes to linger over her coffee and newspaper on weekend mornings. (Many of you may not do this, but just in case.) The two-piece knit or equivalent is more appropriate for leisure hours, morning or evening than a housecoat. This costume also saves embarrassment should a friend drop by unexpectedly. My students told interesting stories.

Whatever you choose as your first garment of the day, let it be the right one. *You* especially should be pleased. Remember you dress for yourself first!

Act II—About 8 a.m.

All Physical Activities for Your Lifestyle Under the Heading Recreation Staples

If you are the person who runs, walks or does other fitness routines before starting all other activities of the day, then a jogging suit and running shoes is the first outfit you wear. Whether you slip into this costume first thing in the morning or any other time of the day, the jogging suit is the proper outfit to wear. This attire is great to wear over your aerobic wear, swimsuit or other workout garments, when going to the gym and back. The especially designed running, jogging or aerobic shoe is appropriate for athletic activities—and you will find its proper use only for this category.

Practical clothes for other morning activities or weekends (gardening, house cleaning, hobbies and activities with children) are colourful, easy-care and washable items: jeans, cotton pants, shorts or corduroy pants worn with all your bright T-shirts. Don't wear them beyond the point of no repair—use them to wash the car. Leave the monster and beer-graphic T-shirts for the kids and adolescents. So when you make a list to purchase T-shirts consider bright colours. I have always said that cotton T-shirts are good for three years—the first year to wear in public, the second year to wear while gardening or washing the car and the third year to wash the car with them.

I hope you like the idea of planning for this part of your lifestyle. These staples—jeans and T-shirts—could be purchased by the half dozen. Three for you and three for your partner.

The sweaters for your Recreation/Staples Category should have substance and be of classic design, e.g., Irish knit, Shetland crewneck or turtleneck. You make the choice. An old nylon sweater will not do; neither will those garments that you plan to "wear out." Would you believe that a man told me he wore his outdated business suits *under* his car while he was repairing the vehicle? Or would you believe somebody would wear her old summer dresses to bed? True stories!

Plan a fun and colourful wardrobe for this category. Don't do the dreary job of housecleaning in an even drearier pair of old pants and shabby sweater.

Every costume in your closet has to be aesthetically pleasing no matter how mundane the "stage".

A windbreaker in waterproof cotton or nylon is the correct jacket to wear over jeans or cords. If you chose these without emblems, stripes or a zipper in a contrasting colour, they are much more versatile to wear for other occasions. A plain red nylon windbreaker fits nicely over your gabardine pants and cashmere sweater when you are traveling. An oilcloth coat that goes below the knee is often more practical than a hip-length jacket. It will keep your knees dry when you walk the dog on a rainy night.

Running shoes with streaking stripes of purple and black are great for sports activities along with your jogging suit. If you are planning to wear runners for other occasions, buy them in plain, neutral shades so that they can be worn for casual travel with your country attire.

Sandals worn with jeans and cotton pants should have heavier soles and wider straps for walking on uneven terrain. Rubber boots can be a real fun purchase when you realize that they come in colours other than black with orange soles. Get red, yellow or any other happy colour. Do not forget to buy a few pairs of colourful socks as well. The colour of socks, shoes and T-shirts do not have to match. Give these staple clothes your signature but leave the jogging suit and shoes for their own stage—sports activities—not shopping.

Act III—The Country/Casual Category

Daily Activities in Residential Areas—Rural Shopping, In-house Meetings, Home Entertainment, Family Restaurants

How to dress for rural shopping and other activities which take you out of your home is a step up from wearing your casual garments. More polish is needed when you step into the stream of community life. Every garment is upgraded from the Recreation to the Country look. I like to think of upgrading rather than to suggest a complete change of style of garments. When you elect to wear a sweatshirt for your Country look, wear it with your pressed denim pants or cords and shoes other than runners, e.g., loafers, sandals or low boots. Shoes really make a difference. This look is still practical but conveys an "I-care-how-I-look" attitude.

The Country/Casual Category is really the most useful one for most people's lifestyles, because it permits you to select garments which work overtime. They are suitable for more than one occasion. The lighter Basic Colour of your chart (**GREY**, **TAUPE**, **CAMEL**) in textured fabric could be used for these often traditional items which are functional and timeless. All fabrics for the Country clothes have the quality of long wearability, i.e., tweed, corduroy, wool flannel, leather, twill and denim.

The traditional blazer in tweed, flannel or denim is an appropriate garment for this part of your life. A wool or leather blouson fits into this category of dressing for the active person and so does a long, belted cardigan. These jackets are usually versatile enough to overlap into the City/Business Category and you could wear them for spectator sports or in transit to the Country.

Hard-wearing pants in classic styling and length are a must for all the different activities in your day and for all the seasons of the year. These pants can be tailored in denim, pressed jeans, corduroy, tweed, leather or whipcord—the last one is used for producing English riding pants. Think of your pants as an investment that will give you great dividends, especially if you are wearing pants most of the time.

When you choose a skirt in durable fabric, also consider it to be practical in style. The skirt needs to allow movement for the action-filled part of your day. A skirt with pleats, or slightly gathered at the waist or an A-line is easy to wear. To complement the skirt and portray the Country look, which is so useful in today's community life, collect some basic sweater styles. You can select them in your bright Accent Colours or in your Basic Colours for a start.

Whether it is a bright red lambs wool V-neck to go over your favourite shirt, or whether it is a Basic Colour crewneck, they are both versatile enough to wear at other times. If you like a turtleneck sweater, choose it in the palest pastel or off-white to brighten your winter wardrobe, or to show off your tan on a cool summer evening by the ocean.

The good looking cotton shirts, blouses or tunics you want to purchase can be in all colours—plain, print or embroidered. Different styles give you the opportunity to display your signature and give the Country Casual look the feel of polish. The duffelcoat is a suitable garment to go with your pants for the cooler climates. It would be fun for the action-oriented person to own such a coat in bright **RED**, or **BLUE**, or **EMERALD-GREEN**. A three-quarter length style is more comfortable for getting in and out of the car, but is often not versatile enough to be worn over a skirt. A raincoat will do double duty for many occasions and needs to be in a basic style and colour. This coat has to be long to cover your skirts and full enough to accommodate heavy sweaters.

Casual shoes which complement these garments are chosen for their comfort as well as for style. Loafers come in several colours and you could make your shoes your personal statement, i.e., always wear bright colours. I suggest you purchase the best quality of shoes your budget allows. The classic sandal with wider straps is one shoe which you could wear with pants or skirts. They will take you many places—even over cobblestone streets. The boot—ankle-high and low heel—has to be of good leather to be of lasting value.

Handbags and belts made of luggage leather, executed in a classic style are a worthwhile purchase. Your accessories should be of quality. Silk and cotton scarves in all sizes, plain or print, are a must in your wardrobe. Jewellery should have some weight to complement sweaters, leather and tweed garments. Bracelets in wood, ivory, jade and handcrafted metals are suitable.

Quality dressing costs money but to collect these suggested garments and accessories over time is rewarding and, in fact, cost effective. The end result is owning a timeless wardrobe for your active lifestyle. These garments also provide a background for those special pieces you have collected and are a good foundation for adding fashion articles.

Act IV—The City

Activities in the heart of any sizeable City—Office, Bank, Shopping, Restaurants, Theatre, Parties

To get dressed for a day at the office or other business establishments, you need clothes in your Basic Colours and garments in the following flat-finished fabrics: gabardine, worsted wool, wool or cotton blends and flannel. Remember the "stage" is one of professionalism and the costume has to convey that. Achieve a well-groomed and polished appearance by wearing garments which relate to the business community in Colour, Texture and Style. The designs have to be updated—an outdated appearance can give the impression that you are not informed and perhaps are not up on the latest in your professional field.

The business suit is the most appropriate costume to convey the message of professionalism. Lapels and straight, padded shoulders communicate preciseness, especially if the fabric is a worsted wool or a gabardine. Jackets which are worn buttoned to the neck convey the same. The closer the suit hugs the body, the more invincible the message.

A tailored jacket without lapels, which has a button at the waist only, or no buttons at all, infers that the wearer is structured but is willing to listen to ideas. Push-up sleeves on business attire are not appropriate unless you are volunteering to work overtime without pay. A fashionable belted cardigan is a good choice for a less structured business environment.

The skirt can also set the mood for what you want to communicate. The straight skirt in worsted wool is very business-like, while the soft crepe or knit skirt conveys a message of flexibility and openness to new ideas.

When you select a sweater for business hours, be aware of the message you are sending. A V-neck sweater over a collared cotton shirt spells efficiency. The same sweater without the shirt but worn with a gold, silver or pearl necklace suggests that you are easy to communicate with. A sweater in a tweedy texture could mean that you are willing to do volunteer work, and you might be approached for it. When you wear a classic crewneck sweater with a small scarf (square knot tied), don't be surprised when you are asked to run an errand—you look so sporty and fit. A large silk scarf draped around your neck and high-heel shoes will deter this kind of action message.

A dress conveys the same message as a suit. The structured garment with design angles (pointed lapels, or stand-up collar) communicates the exacting nature of business. When the lines of dress become more relaxed it conveys an openness to interact with people.

When you choose your business attire, take into consideration which part of your garment is most visible. From shoulder to waist is usually the most noticeable part which can give the impression you want. When you are on your feet a fluid skirt might soften the message while pants add precision to the costume.

Pants for business attire, classic in cut and in a fabric that holds a crease well, are most appropriate for the action-oriented person. Pants for the city environment need special consideration, particularly for the woman who is starting up the corporate ladder. To make pants acceptable pay extra attention to the blouse or shirt and jacket of your costume. City pants are not worn for comfort alone; they must convey a message of preciseness.

The shoes for the city costume are dressy, medium to high-heel, with thin soles (contrary to crepe soles). Shoes have to be highly polished and of the best quality you can afford; they say "class." Sandals are correct for the summer and so is a dressy (not outdoor) boot for the winter.

To accessorize with a good bag is a smart investment. It need not be an attache case or briefcase, but it should sit well on a desk or table. Jewellery that gives your costume your signature without interfering with business at hand, includes: beautiful earrings, necklaces, bracelets (not klunky) and rings which convey that you paid attention to your costume. Remember, anything that looks like a costume for "play" is not for the city. No jeans, tank tops, see through blouses, or gauze skirts. (How and why Casual Friday ever came into being, I have no idea.) Clothes that are too bright, or too flashy or revealing don't have a place in the business environment.

Dressing for downtown shopping, having lunch with a friend, or attending to some business, requires a costume that can be like the one I described for business, but you might want to add more jewellery or choose less practical colours. "Dress for a successful life" rather than "Dress for Success" should be your motto. You will enjoy the feeling which comes from being well-dressed. (The better you dress, the more attention and better service you will receive.) Any meeting, whether community, fundraising or other, can be upgraded by attending them in business-like dress.

You will also need a coat, cape or wrap, depending on your personality, to fit over all your clothes of different lengths and colours. The most useful coat for rainy climates as well as for traveling, is the raincoat which is described in the Country/Casual Category. Some climates do not warrant a raincoat and you will need a seasonal coat—consideration for style is the same as for a raincoat.

Act V—The Dinner Hour

This part of the program deals with the dinner hour. I would like you to agree that depending on where you eat, a change of costume from Act II is warranted. A different stage setting demands a change of costume. Please take time to upgrade your dress for the casual dinner at home which centers around preparing the meal and being with your family. The costume which often has not changed since Act II (8 a.m.) is now going to be polished. Freshen your make-up, add a splash of cologne and change into a clean top. This takes very little effort and rejuvenates you and your family. You and your loved ones will reap the benefits from having created a pleasant atmosphere.

The career person might appreciate putting on something soft and unconstructed (no, not the housecoat) when she comes home from the office. A two piece in cotton-knit will give you the feeling of change of pace (even if you are the family cook). Knit pants (no stripes on the side please) with a matching tunic and gold or silver sandals provides a touch of glamour. Accessorize your cotton-knit calf-length full skirt with a matching T-shirt and a beautiful belt. Some people might choose a caftan or an ethnic full-length dress. Wear pretty slippers in matching colour (or gold/silver) for your costume; wear anything that makes you feel pretty and special. Give yourself a lift after a day at work away from home. It does not take much time and you are worth it. Before somebody protests and talks about cleaning bills, please note that our costumes for dinners at home, are washable. Some of my students have told me that being able to use washable garments gave them the incentive to get out of their jogging suits or jeans and dress for the family dinner hour.

For every area of your life, every time of the day, I call your attention to the aesthetics. I have a challenge with people who wear clothes only to cover their bodies. I hope you are open to having some fun while creating your costumes making you the best dressed, main character on the stage of your life.

Dressing for special company at your home. "Special" is anyone with whom you care to share a meal. An evening set aside for entertaining friends is an event. Do not make in a non-event by underdressing. You want to convey a caring attitude towards your guests by the way you dress as well as with your surroundings. The secret is to combine practicality (unless you have domestic help or a caterer) with glamour. I want you to feel good about your costume.

A long skirt (calf or ankle-length) or a pair of soft pants worn with a blouse or an evening sweater are pretty and appropriate outfits. The sweater may be made of lurex and have a flattering, lower neckline. Long, easy fitting dresses—maybe in linen for summer and knit for winter—(not evening gowns) in beautiful tones to complement the colour scheme of your living room are another possibility. I have never quite understood how anybody knowingly could wear a hot pink dress in an orange living room.

Most important to your dinner attire at home are the accessories—earrings, short necklaces, pins—anything which does not interfere with cooking or serving. Do not under-accessorize and give special attention to your shoes. A soft, low heel, leather shoe is appropriate (don't wear house slippers).

The categories of how to dress for dinner at home assume that you are the main character on your "stage" who sets the tone for your family and guests and makes the evening a special occasion.

Dressing for dinner at a family restaurant with or without children, takes little effort, but some consideration for your appearance is required none-the-less. You are still the one who sets the level of dress and behavior at your table regardless of how casual the atmosphere may be.

Pants or skirt (even in denim) are appropriate as long as you wear them according to the Country Casual look with washable shirts and blouses, or even a special T-shirt (no logo). The shoes are the one accessory that can take this look from Recreation to Country even though they are under the table most of the time. Do not forget to add some jewellery, earrings,—anything which makes you feel as though the evening is an event, no matter how casual the surroundings.

The children might like to wear a shirt instead of the T-shirt they wore all day, and the girls may choose a dress or pants. I am sure you will appreciate their costumes which convey an "I-care" attitude. These small efforts will also help to shift your busy thoughts of the day to a more relaxed time of sharing a meal with others.

Dinner with friends at their home spells fun and relaxation. To contribute to the evening and to show your gratitude, arrive in a costume which conveys appreciation and respect. Don't forget to bring a smile, flowers or a bottle of wine. If your hostess/host is the casual type and you know she/he is going to be in jeans, wear pants also, but this is where the resemblance stops. Wear your prettiest blouse or sweater, feminine shoes and your best jewellery. Consider your hostess always, but do not dress down.

Dinner with friends who make it an event is fun. Your most unconstructed, flowing dress in silk or silk-like fabric is perfect for this kind of an evening. Silk pants with a blouse or a sweater and lots of jewellery is another way to look right for this event. If you have blouses, skirts or pants which are your special signature, e.g., ethnic or pieces of art, wear them; they add to the atmosphere. Wear your prettiest high-heel shoes with sheer stockings, special low sandals or ballet slippers, and jewellery from the safety deposit box for the occasion. Why wait for a grand occasion to wear your best when an evening with friends is so special?

I was told of dinner party given by a woman for her close friends in her elegant home. She engaged a pianist and cellist for this special evening; but two of her male friends arrived in jeans and running shoes for this elegant affair, while their spouses were elegantly dressed. Not appropriate, is it?

When you choose your costume, wouldn't you consider your friends' surroundings? The colour you have chosen for their dinner party may be your favorite one—that in itself will make you feel happy, but it certainly helps to make an impression of ease when you are in harmony with their "stage." For instance: your friends' living room is done in **APRICOT**, **SOFT GREEN** and **GOLD**. One of your favourite dresses is **HOT PINK**. Of course you are an outgoing personality and you are certainly going to contribute some interesting stories to the evening, but couldn't you do it in an oyster-white silk pantsuit with some extra special accessories? The hot pink dress will be great for an evening for two in a restaurant, but will clash horribly in your friends' living room. You might think this is going too far, but think about it.

Dinner for two. Dinner with your husband, lover or friend warrants a plan. Do not ever make it a non-event. Visualize yourself at the table in an intimate restaurant. Your dinner partner sees you from the waist up. Don't you agree that the emphasis of your dress should be given to the top part of you, including hair and make up? The upper part of your costume, whether it is a dress, a blouse or a sweater, has to be of a structure and style in which you feel at ease. If you have to correct a low neckline or shawl all through dinner, part of your physical energy and your attention is consumed. You might be better off wearing something closer to the body and add some extra jewellery around your neck. A plain neckline shows off some great earrings, or you may want to drape a silk or chiffon scarf around your neck. Can you handle it? A satin shirt is very becoming especially by candlelight. When you get up, you will be glad that you wore a skirt which had a bit of fullness or drape over your hips and tummy rather than clinging to your body. Unless, of course, you like a tight waistband so you won't overeat.

High-heel shoes or sandals with sheer pantyhose complete this costume. Do not carry a large handbag; it looks as though you brought your own bread and salami.

Act VI—An Evening Out

At the Theatre or the Opera

Every day is a play. What are you going to do today? What is your part? That is the fun in dressing. An evening to enjoy the atmosphere of the theatre, opera or a musical (especially opening night) is asking for you to dress in something special. Forget that somebody told you that it does not matter what you will wear as long as you are interested in going to the theatre or opera at all. Create your own theatre within the theatre. Succumb to the surroundings and set the tone within yourself to experience a wonderful evening taking in a performance of which you become a part.

Your most beautiful silk dress in your favourite colour is just the start. Let the style be unconstructed and flowing to set the mood for you. You should float not march. Remember the difference in stride when you wear runners or feminine high-heel sandals? Do you have that special feeling when you wear dressy shoes? You straighten up and have more poise; you feel more elegant. That is the feeling I would like you to portray when you're in special surroundings attending special events. Please do not underdress. You can wear a skirt or pant with a brocade jacket, an evening suit with rhinestone buttons or your prettiest blouse with narrow pants and high-heel sandals. Wear your best jewellery and give your make-up and hair some special attention as well.

The main consideration should be that whatever you wear is an upgraded look from what you would wear to the office. If you have no time to change from your business suit into an evening outfit, it is just a matter of being prepared. A switch to sheer pantyhose and higher-heel shoes does wonders. Exchange the shirt or blouse you wore with your suit for a camisole. Then add some interesting earrings and drape a large shawl over your shoulder for a special effect. Every occasion can be made into an event by you—do not let the moments pass.

Act VII —Midnight

The day has comes to an end, and I would like to talk about your midnight apparel. Whether to wear pajamas, nighties—short or long—is a very personal choice. I hear the logo T-shirts have even made it into the bedroom. Really? You might like pajamas for one part of your life, or a nightie with matching negligee for another. Plan for it. Always ensure that the colour is in harmony with your surroundings. (You really do not want to be lulled to sleep on a cruise in your plaid flannelette pajamas, do you?) Do not forget to wear your special perfume at night.

Dressing for Special Occasions

Dressing for special occasions and venues shows an "I-care" attitude. A little thoughtfulness goes a long way to make an event special and fun. Are you and your clothes in harmony with your surroundings? The costume designer in me always looks for opportunities to project the right image. No, it is not dress-up time, just some common sense and fun advice. It is easy! A blazer for the yacht club lunch, a pastel or off-white charmeuse blouse with black silk crepe pants for a Mandarin restaurant—not the other way around. Maybe a longer skirt with an embroidered blouse and sandals to wear to a Mexican restaurant. He could wear his banded no-collar shirt and beige or black slacks. Would you not feel silly in a navy blue suit? Or black? I would not suggest you run out and buy a "costume" for every restaurant you go to, but just think of yourself in harmony with your surroundings. Have fun planning.

How about you at a special opening at an art gallery? For a Dale Chihuly glass display wear clear bright colours in flowing fabrics, i.e., a blue organza coat over white silk pants and pink camisole with lots of coloured glass beads. I like the idea of a forest green handwoven chenille stole, jade and silver handcrafted jewellery with your grey flannel pantsuit or taupe knitted dress for a showing of West Coast art. For a November evening reception to view antique Flemish paintings wear a deep purple or garnet-red velvet jacket over a brown or black skirt or with a pair of pants. Add lots of gold jewellery.

I would have loved to own the jacket which a guest wore to the opening of Alexandra Nechita paintings which we attended.The lady had on a brightly abstract-painted silk jacket which looked just like one of the canvasses on the wall. Look into your closet and see what you can find for evenings like this, because a black or navy suit might mistakenly identify you as a buyer at an art show. Dressing for the venue shows your "I care" attitude.

When we arrived in Venice on the Orient Express from London we agreed that this was certainly "the only way" to travel. The trip was beautiful, the food and service excellent and the ambiance—oh, the ambiance. The brochures had stated correctly "…you can not be overdressed on the Orient Express." True, indeed!

All male passengers wore tuxedos for dinner and the ladies outdid themselves. There were beaded chiffon and satin gowns, feathers and glitter everywhere—most of them resembled the style of the twenties. My husband wore his navy blue tuxedo. I wore a pale beige chiffon dress completely covered with silver, gold and white beads. We dressed for breakfast, we dressed for lunch, we dressed for afternoon tea. We have never traveled with so many clothes ever. We sent them all home before we continued our travels.

This was a very special trip and you must agree that the effort to dress in harmony with your surroundings on the part of all the passengers made it a special event and experience for all of us. Before we went on this trip we saw a travelogue about another beautiful Orient Express train bound for Istanbul. The passengers on that train seemed to have had a great time in jeans and cowboy hats partying at the bar. Now why would anyone want to do that? Which one of the trips would you rather take? The price is the same.

I have in front of me an invitation to a California wine country feast and art show. The invitation reads: "Dress with a splash of colour!" It will not be a costume party, but an event where guests are asked to give consideration to the colours they could wear. Dress in harmony with your surroundings and remember these are the little things which count in life.

My husband and I especially enjoy concerts at the Music Academy in our city which are given at 10:30 in the morning. June, the founder and artistic director of the concerts, sets the stage and ambiance for those most enjoyable mornings. She herself is beautifully dressed and there is always a large basket of flowers on the stage and at the entrance. Refreshments are served in flower decorated baskets. Being there, surrounded by well-dressed men and women of all ages in a lovely setting with good music is an enjoyable experience. It gives me sincere pleasure to see that the audience takes pride in contributing to the elegant atmosphere, even though it is not an evening performance—no jeans and running shoes here.

Operas, concerts, ballets and other cultural events warrant respect for the artists and the environment. It is the audience and the ambiance which support the performance to make the occasion a memorable one.

For many invitations you receive, you can wear your "little black dress" of course. Somebody said that anybody at any age can wear it. I agree. Anybody could, but why? Would it not be more exciting to make your own statement? Visualize a silver-grey flannel suit with rhinestone buttons, satin collar and cuffs or my **TAUPE** friend's costume: a beautiful antique lace T-shirt in dark taupe, (only the **TAUPE/COOL-BROWN** personality would have it in her closet) an above-the-ankle crepe de chine skirt in the same colour and lots of pearls. Our **PALE BASICS** will make a statement just by the colour alone and stand out in a mostly **BLACK** crowd. But if you want to wear **BLACK** for an evening event, pick the most glamorous fabric and style—the choices are endless.

*I will wear my true **NAVY** costume: a pleated navy blue chiffon skirt, above the ankle, a navy charmeuse top and a fitted V-neck vest. Here is where the **NAVY** look stops. My skirt is not lined (I wear opaque hose). The vest is covered in sequins and red satin pumps are on my feet.*

The man in your life can definitely add to the atmosphere by dressing appropriately and with an "I-care" attitude. It is easy when you get an invitation with a dress code that reads: "Business suit, Black or White tie." You know what style of dress you will wear and so does your partner. When the invitation is for a special evening, and you know that the hostess has decorated her home for her guests and the table is set with care—be it for a birthday, Christmas or other occasions—the first question that comes to a woman's mind is: What am I going to wear? The husband already decided to wear a sweater and sportshirt, no tie, no jacket. I'm sure the hostess would appreciate that her male guest wear a jacket and a tie, acknowledging the efforts that have gone into the preparation for the evening. If that is too much of a challenge, then a jacket with a sportshirt or a sweater with shirt and tie is acceptable. I wonder what we all would wear—women and men—if the invitation called for casual dress?

Here's my answer. By now you are familiar with the wardrobe plan for different venues. For the invitation that calls for casual dress chose your garments from the Country/Casual Category chart and not from the Recreation/Staples

Category—unless the event is a picnic, maybe.

The easiest way to respond to the call for casual dress is to have one part of your costume structured, e.g., wear loafers instead of runners (here I go again!) and put on a pair of pressed jeans or chose twill pants. Top the pants with a blouse or shirt, but do not reach for a T-shirt. Remember, just one garment with polish immediately elevates the look from Recreation/Staples to the Country/Casual Category.

Today's casual approach to dressing is called "High-Low" dressing. One part of your outfit comes from a catalogue or from one of the many specialty stores where you can purchase items of clothing for under a hundred dollars. The "High" part is your designer accessory—Gucci shoes, Epi bag or Armani jacket?

I always add some polish to my casual outfits. My favorites are my Lands End beige twill pants, a silk or linen blouse for the summer and a cashmere sweater in the fall with quality loafers and belt and my signature bracelet. This "High-Low" image is simple, smart and current. Whether you wear pants or a skirt it portrays an "I-care" attitude for a "casual" dinner invitation.

Body Image

Most women are critical about their bodies. They invent imaginary flaws and exaggerate real ones. This part deals with the physical body and how to dress to enhance the best features and to camouflage the weaker ones.

Design Lines

If you know the proper design lines, your costume will give you the added confidence to purchase the right clothes for your figure. Clothes designed to enhance your figure and diminish figure problems enables you to wear fashions which are becoming to you. Once you understand the camouflaging effect certain clothes can produce, you can go ahead and experiment and discover what works best for you.

- Vertical lines cut width and give height.
- Horizontal lines make a figure appear wider and shorter.

By considering even the smallest detail in designs, and the use of colour you will have fewer mistakes stored in your closet.

To minimize the look of a along neck, **think horizontal:**

- Wear heavy, constructed jewellery at the base of your neck.
- Wear a high turtleneck.
- Wear a scarf tied close to the neck with a knot or bow.

To give the short neck maximum exposure, **think vertical:**

- Choose a neckline cut away from the neck and wear long strands of pearls or a chain necklace.
- Wear a V-neck sweater or V-shaped neckline, e.g., lapels.
- Wear a scarf tied low.

To minimize broad shoulders, **use vertical** construction lines:

- Raglan sleeves
- Bust contoured from shoulder seam
- Square inset sleeves

To add width to your upper body to balance the lower body, **use horizontal** construction lines:

- Choose a yoke and fullness on the crown of the sleeve.
- Choose dolman sleeves with padding or epaulets.
- Choose tops with a dropped topstitched shoulder seam.

To minimize the look of a long torso, **think horizontal:**

- Choose a yoke to cut length.
- Wear a sweater belted (5+ cm/2" belt) at waistline.
- Choose a shirt with contrasting colour, skirt and belt in same colour.

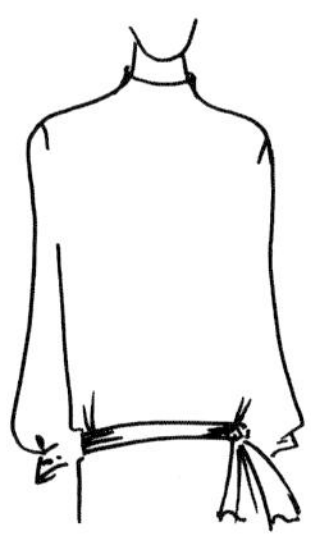

To lengthen short upper body, **think vertical:**

- Wear an overblouse that rests on your hip bone.
- Wear a sweater belted (5 cm/2" belt or less) below the waist, resting on the hip bone.
- Choose a skirt with a loose waistband and blouse the top below the waist. Choose the top and belt in same colour.

To minimize waistline, **use:**

- Contrasting colour or fabric inset on belt (18-20 cm/7-8" wide)
- Belts and scarf tied or buckled at side
- Belt loops (2 cm/¾" wide) break width of waist through vertical lines

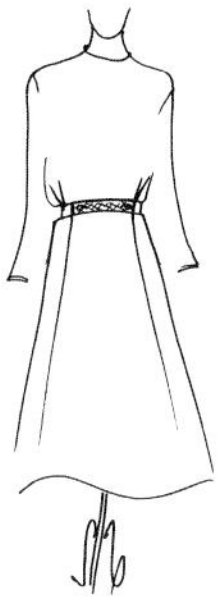

To minimize hipline, **think vertical:**

- Six-gored skirt or side seams moved (5-6cm/2-2¼") from side towards front
- Four-gored, center front seam
- Jeans-cut center front opening and pockets minimize a heavy waist.

Jacket Proportions for Different Figures:

1. Waist length—no camouflage for hips and legs. Best suited for figures with 92 cm /36" hip measurement and small bust. Easier for larger figure when worn with a skirt.
2. Hipbone length—good for shorter women. When hips and thighs need no camouflage, a good jacket with pants as well as skirts.
3. Crotch-length—camouflages and gives length to legs as well. Good as a basic style with pants or straight skirts.
4. Thigh-length—good camouflage for body. Best for 167 cm/5'6" and taller but wearable for smaller women if same colour as skirt.
5. Walking coat or knee-length—covers well but is really best suited for taller women. (For shorter women, do not wear in contrasting colour).

Closet Organization

It is interesting that through the physical structure and organization of a closet the owner also reveals their personality and lifestyle. Being well organized is a **NAVY** trait and if you fall into the **BLUE** category you will love this chapter. The organization of the closet is definitely the **NAVY'S** paradise and the **BROWN'S** nightmare.

The **NAVY** personality likes wearing clothes which are primarily made of hard-finished fabrics. Shirts have proper collars and cuffs, the jacket has lapels and good shoulder structure, and very important, **NAVY** personalities dislike wrinkled garments. For that reason they want lots of hanging space rather than drawers, and they want sturdy hangers for their jackets, and special hangers for skirts and slacks.

The **BROWN** personality, on the other hand, wears clothes that are less constructed and made of fabrics that have texture—with preference for cords, sweaters and leather garments. For this reason it is appropriate to have more drawer space than hanging space. Sweaters and knits as well as leather garments should not be on hangers (except for leather jackets and coats), therefore open shelves and drawers behind doors are necessary. A large open basket is handy for dropping clothes.

The **BLACK** and **TAUPE** (**COOL-BROWN**) personalities prefer wall space on which to display their choice items of clothing as well as enough room to hang them neatly, or to store them folded in their proper place. These personalities derive pleasure from seeing their collectibles as decor rather than hiding them. A whole wall could be fitted with a brass rod on which to hang belts, necklaces and scarves. A special garment (e.g., a kimono) placed on a brass hanger becomes a stunning wall decor.

BEIGE, **STONE** and **GREY**, being lighter personalities, have the tendency to be open to all possibilities for organizing a closet or dressing room but have a distinct affinity for the characteristics of their Parent Colour, i.e., **BEIGE** to **BROWN**, **STONE** to **BLACK** and **GREY** to **NAVY**.

After having decided on your colour group, clean out your closet, no matter what its condition at the time, in order to derive the maximum benefit from the guidelines in this book. When you open your closet door you will discover, along with thousands of other men and women, that you have too many clothes and nothing to wear. The reason for the accumulation of clothes is mostly due to a mixture of colours from different palettes and the reluctance for weeding out things from time to time. To get started, set a date in the very near future for organizing your closet.

How to Organize Your Closet

1. Remove all clothes which are on hangers and separate them into categories of: spring and summer, fall and winter, mending/alterations, cleaning and not worn for two years.
2. Remove all shoes and separate them into categories of: spring and summer, fall and winter, polish, repair and pinch (these were bargains?).
3. Return your garments to a clean, empty closet—hang up only that part of your wardrobe that pertains to the current season. Arrange garments according to colour. If you have double racks for blouses and slacks, one over the other, arrange them in colour groups as well: hang your blue shirt on the top rod directly over blue pants beneath.
4. Return shoes—place shoes on racks or low shelves, grouped according to colour, directly under garments of same colour. Disregard shoe type.

When all your clothes for the current season are back in your closet, you will see columns of colour appearing (a sight which becomes more established over the years). Most probably you will discover that "your" colour is the dominant one. If not, don't worry about it at this time. The two things you will experience, however, are order and more space in your closet. Most husbands believe that a full closet does not justify the purchase of more clothes. Wouldn't you like to show him your empty closet now?

5. Return bags and hats—the top shelf above the rod in the bedroom and

hall closets provides storage space for bags as most women can only reach the front edge. Clean and polish up-to-date handbags before returning them to storage. Place the handbags in their proper colour slot above your garments. Store hats in boxes at the back of the shelf or slip them into clear plastic bags and pin them to the wall above the shelf with a thumbtack.

The first step organizing your closet is now completed. It is suddenly colour-coordinated, more spacious and only the current season's clothes are in sight. The next step is to:

6. Take shoes to shoe repair, take soiled clothing to the dry cleaner or laundry and put those needing mending or alterations in a place, ready to be fixed.
7. All off-season clothes go into hiding for the next few months, but should be washed, cleaned or repaired and placed on good hangers before storing. It is a pleasure if everything is ready to wear for the next season at the time you rearrange the contents of your closet. Store off-season clothes in another closet that has also been cleaned. If another closet is not available, use white opaque garment bags and make room in your hall closet, attic or spare room. If all else fails, store the garment bags in your one and only closet.

Now there is only the pile of garments left that you have not worn in the last two (or is it ten?) years. However, you will be adding more to this pile after you have organized your drawers and shelf space, so don't worry about it now. The next step is to:

8. Clean out all drawers and shelf space in your bedroom, bathroom and other personal surroundings. Line drawers and shelves with scented shelf paper. If you are reading this before cleaning your closet, add scented shelf paper to your shopping list. Scented shelf paper is available in a variety of pleasant fragrances suited for both men and women. It is pleasant to open a drawer which is colour-coordinated and has a lovely scent.

If you need containers for your off-scason clothes, sweaters, bathing suits, gloves, sunglasses, scarves, ties, shawls, underwear, etc., let me suggest using baskets as organizers. Get ready to buy baskets in all shapes and sizes because you will need several. This is the fun part of cleaning your closet. Baskets are practical as well as aesthetically pleasing. It is great fun decorating with them, and nobody knows that they are the hiding place for your ski socks or bathing suits. The basket that is constructed to resemble a suitcase is the most useful for storing sweaters in plastic bags to which you add some herb moth-proofing. By storing your off-season clothes in places other than your dresser or on your shelves, you will have more space for your present wardrobe and have everything within easy reach.

Fold sweaters and arrange by colour.

Divide all clothing into seasons.

Scarves, ties, shawls and gloves do not have to be separated by season and you can keep all colours. Square, flat baskets are handy for keeping your accessories organized when they go back on a shelf or in a drawer.

Discard your undergarments that no longer serve their purpose. You know the ones I mean: stretched-out waistbands, faded colours, frayed edges. Make the purchase of new underwear a priority on your shopping list. Visualize yourself in something pretty, dainty and feminine—treat yourself. Fold your fine undergarments into squares or roll them up neatly and place a sachet with each pile of underwear.

Night gowns and pajamas should all be in one drawer. Keep off-season pieces in pretty lingerie bags made of cotton and lace. I use these same lingerie bags when traveling.

Pantyhose, stockings and socks need to be inspected. Discard all those with holes, snags or runs. Separate pantyhose with reinforced toes, coloured pantyhose and plain sheers. Place each into a small plastic bag and colour code each package with coloured labels or dots. It is a pleasure to open your stocking drawer first thing in the morning and can lay your hands on the right kind of stockings immediately. If you are **BROWN**, not to end up with a drawer full of empty plastic bags at the end of the month is the challenge. Actually I am quite pleased that you are still with me—and maybe you even find organizating your closet fun?

All of your hanging, drawer and shelf space has now been reorganized and your off-season clothes are clean and ready for you to wear when the time comes. You have used scented paper for all your drawers and shelves, and your personal belongings are stashed neatly in baskets or other organizers. Most of all, everything is organized according to colour.

The only clothes remaining for your attention are those things you have not worn in the last two years or more. Do you like to save things? This is the time when your true personality surfaces.

NAVY: If you are a saver and a sewer, you may want to take the garment apart and keep the larger pieces (skirts, sleeves, backs) and save the useable fabric. Wash, clean, press those items and store them in a plastic bag. Maybe you will find a use for them another time.

COOL-BROWN: If you are sentimental about an outfit you can store it in a special box lined with pastel-coloured tissue paper, tie it with a pretty ribbon and place in on the top shelf of your closet or in a trunk in the attic. DO NOT put it back in your closet.

BROWN: If you plan to lose weight and hope to fit into a smaller size again, don't keep those clothes. When you in fact lose weight, reward yourself with a new outfit in an up-to-date style.

BLACK: If you are a **BLACK** personality you will probably like to make room for new things in your wardrobe and give away the old.

You made it! Now, everything in your closet is a pleasure to behold. You have made progress toward becoming a very organized person, even if it did not come easily at first. Just think how much time, energy and money it will save to have a well organized closet.

Gems from my Collection

Closet Organization

If you have **NAVY** *traits in your personality like my dear husband, then you will probably hang on to things in your closet "just in case." Francis showed me a box before our seventh move, and asked whether he should keep its contents. When he took off the lid it revealed the Halloween outfit he wore when he was seven years old. It is a really cute costume: plaid shirt, leather chaps and cuffs, cowboy hat, bandanna, a holster and that all important six-shooter—all wrapped in tissue paper. My husband likes to collect and keep things, but everything is in order and in its proper place. If you are married to a* **NAVY** *man don't ever think of throwing things out. Keep them in boxes neatly tied with a bow.*

How to create storage space

Here is a tip to create more storage space which I have used successfully in our own home as well as those of my clients. Take 12" to 14" off one length of wall in any room and install lots of shelves behind floor-to-ceiling mirrored sliding doors. I did that in our townhouse. The living room and dining room flowed together and we converted one of the dining room walls into a storage space for china, glasses, table linens, wrapping paper and off-season clothes. By doing that we gained a lot of space and visually, the room appeared larger due to the mirrored wall.

Remember to review this chaper every six months

Guidelines for Building a Wardrobe

When you have taken stock of your current wardrobe and have completed your wardrobe plan chart, you will have discovered all the possible outfits you can get from your existing wardrobe. Decide what additions will work with what you already own. Make a list of the garments needed. Set aside two blocks of time for purchasing the missing items. One time is for browsing and scouting, and the other is for shopping. I know it is often difficult to find the extra time to look for things before shopping, but it is worth your while when you consider what you want to accomplish.

Check magazines and newspaper ads for inspiration and to become informed about new trends. Leaf through the magazine, advertising included, and look at everything from hairstyle to neckline to shoulders to hips to feet. Study, analyze and compare. Do not get discouraged or even depressed when you see high fashion to which you can not relate. Do not feel you have to wear any one of the designs featured. Continue to analyze the display of blouses, dresses and the fit and length of jackets. Check out the skirt length and style for the current season. Since you are secure enough in your own style, you do not have to adopt every fashion fad which comes along. Examine the jewellery, bags, eyewear and how the scarves are tied. Scrutinize each piece of clothing for what it has to offer.

The results of your exploration will be that you know the current fashion trends and that you will be able to select your basic clothes for your starter wardrobe in classic design, but in updated styles. You might also have discovered some styles, details or accessories which will help you in creating your own signature. The starting point to building your wardrobe is right at home. The more information you have about the items on your shopping list, the more successfully you will build your wardrobe.

Here are a few more tips to remember:

Your raincoat

- This coat will stay current longer if it has no waist seam.
- A buttoned back-slit—if it has one—makes it more useable for a housecoat on trips.
- Leave the coat the longest length you can sensibly wear.

Other coverings

- Any coat without a waist seam remains fashionable longer.
- Capes and shawls make a personal statements and are timeless.

Jackets

- Check the size of armholes. They should be able to accommodate sweaters and full-sleeved blouses.
- Wider shoulders balance hip measurement.
- Single-breasted styles are more versatile and slimming—excellent for your travel wardrobe.
- Hipbanding on blouson jackets should be 5 cm/2" larger than hip measurement.
- Purchase a belted cardigan or knit jacket a size larger than your normal size.
- The windbreaker should be as plain in design as possible, without losing the style, to overlap Country/Casual and Recreation/Staples Categories. This garment is also useful for travelling.

Skirts

- The basic difference between City and Country fashion is practicality and ease of movement of skirts.
- Denim as a fabric has become a classic and is appropriate for many occasions except for corporate business attire. To invest in a good denim skirt and jacket earns long-term dividends. Denim is not a colour in my theory but a statement.
- A full skirt becomes even more comfortable on a long journey when the

waistband is elasticized—at least in the back of the skirt.

- Culottes are an alternative for a skirt in the Country/Casual Category, and are wonderful for a weekend get-away. When made out of gabardine in your Basic Colour they can take you to a board meeting when worn with city accessories: silk blouses, tailored jackets and thin-soled shoes. Great for travel!
- Hard-wearing twill, gabardine or flannel are the best choice for your basic wardrobe.

Pants

- Shop for basic style and fabric for your first pair of pants.
- Wear them like a bare canvas against which to display your signature blouses, shirts or sweaters.
- Full length pants should touch the ground at the heel when barefoot.
- Pants with a zipper in the side pocket are more flattering to a mature woman's body than the front zipper styles.
- Cords and knits should not have a pressed crease.
- Do you work or live in jeans? If you live in jeans, buy designer jeans in updated styles, or give the jeans a press.
- Bermudas and shorts can fit into your closet if they suit your personality and body. They are good alternatives for country travel and, of course, for the Recreation/Staples Category.

Sweaters

- The three basic sweater styles in the wardrobe plan are timeless, but watch for slight variations of fullness and length during different fashion seasons. If you prefer a looser style, purchase your sweaters one size larger.
- The turtleneck sweater is preferred without a zipper closing. The neckline should be loose enough to run a finger easily inside to ensure it is fashionable. The turtleneck T-shirt rides close to the neck for practical reasons.

Blouses

- Cotton shirts and blouses can become collector's items and your personal signature.

Look for blouses in specialty stores

- Men's shops: tuxedo shirts.
- Boutiques: ethnic clothing—Mexican, Indian, Austrian, etc., are good sources for signature blouses as well.
- Chinatown: silk and embroidered blouses.
- Lingerie departments: nightshirts or men's pajama tops.

Dresses

- Versatility is the key word for dresses.
- Purchase the coat dress in a larger size to accommodate blouses and sweaters—use it as a coat over a silk two-piece dress.
- The knit dress should be large enough for your hips to move easily—it should not cling to your body; it is more slimming that way.
- The two-piece dress in silk, wool, cotton or knit is your best purchase.

For evening

- Invest your money where your lifestyle is.
- Do purchase a two-piece silk dress.
- Accessorize a silk shirt.
- Look for evening gowns in bridal shops if you have the occasions to wear them.
- If you have worn an evening dress for several formal functions, shorten it to dinner-length to give it another reason to be in your closet.

At home wear

- Consider the colour of your living room.
- Consider your physical need of colour, e.g., bright for a lift or pastel for calmness.
- Let it be washable.
- Skirt and pants length should not be longer than you can wear in stocking feet.
- Look for pretty and comfortable shoes.

Bags

- The large carryall should have room for holding a small clutch bag.
- Consider the contents your bag must hold before purchasing it.
- Your bag for the city should be able to sit well on a desk or table.

Shoes

- Wooden, stacked-heel shoes are practical.
- For longer wearability, consider having your new shoes half-soled before wearing.
- A low walking shoe in a more feminine style to wear with pants or a skirt is the most versatile shoe with which to travel.

All Staples—Recreation Wear

- Purchase these in the brightest colours for action; soft shades for relaxation.
- Order them through a catalogue, like Lands' End.
- Purchase two jogging suits at a time to mix colours of top and bottom.
- Plan to have brightly coloured socks.
- Several cotton scarves are a must.
- Buy cotton T-shirts in the best quality cotton and in a style so that they can be worn in the city along with jewellery and under your tailored jacket. Add shoulder pads if you like.
- A matching bathing suit coverup looks like a "cover up." Try a cotton shirt, a big sheer top or a sarong instead.

Lingerie

- Keep labels from previous garments for re-order.
- If you can't find slips for your height, buy a long one and hem it to desired length.
- Choose white or colours which match your outer garment.
- Purchase natural-coloured undergarments to wear with white outer garments.
- Purchase bright colours for your action wear.
- Match your sleepwear to your bedroom colour.

- Consider the colour of your kitchen when purchasing your housecoat.
- Match your bathrobe to your bathroom.
- Consider camisoles to wear under a jacket—but don't take off the jacket.

Tips for Sewers

When you have chosen a colour group for your wardrobe, consider buying your wool fabric in large yardage. You most likely thought of making the jacket, skirt and pants in the same fabric. Maybe you would like to take this plan one step further? Make two skirts, one straight, one full. Keeping fabric on hand to duplicate a pair of pants or to create a skirt in an updated design is another practical alternative. Remember that the jacket took the most time to sew, and should last three years or longer. You might also have invested in a cashmere sweater in just the right shade to match your pants. The pants will most likely wear out sooner than your expensive cashmere.

When you purchase materials, silk, cotton, etc., plan to buy enough to make extras like scarves, camisoles, hankies or lining for the jacket. Even a matching bra and panties to go with your silk shirt would be fun to create. Just don't make your daughter a matching dress (or a shirt for your husband).

Carry fabric swatches, neatly cut, in a plastic credit card hold for easy reference, along with fabric requirements for each garment.

THE BROWSING DAY

WITH ALL THE INFORMATION GATHERED IN THESE CHAPTERS YOU will now be able to have fun browsing wherever you are and then be able to shop with assurance and purpose. You have accumulate the knowledge you need to plan your wardrobe. Spend this day of browsing by yourself. Don't take along a friend! You need to be able to concentrate on what it is you want to accomplish. Dress in something comfortable and stylish. On that day wear attractive undergarments, a well fitting bra and shoes that keep you walking in comfort.

Remember the temperature level in the stores and shopping malls. You might do well to check your coat and umbrella. All you want to carry is a small shoulder bag and your notebook. Relax while you are walking through the stores. Look only at departments which carry the items on your list. Do not get side tracked by looking at displays of table linens and fur coats. Survey only what is on your list.

First, look for the most important item on your list. The jacket is so vital for your wardrobe that you should be fresh and willing to try on different styles without feeling pressured. Shoes are also easier to try on in the morning than at the end of your browsing day. Give these two items your immediate attention.

When you are looking for skirts, pants or dresses, take more than one garment into the fitting room. Take your time. You are just looking and are not making a decision about anything today. Try to have fun! Do not get discouraged by the dreadful lighting in the dressing room. Smile! If you think that one or more of the items you tried on might interest you, ask the sales person to keep them for you until the next day. Ask for her name and give her yours. Collect data and take notes as you walk through the different departments and stores. Write everything down including the store, phone number, salesperson's name and list the items you might be interested in purchasing—description, size, price and maybe even the stock number. Collect information with commitment to purchase. Act professionally and let the salesperson know that you are scouting today for your purchase tomorrow.

Take about two and a half hours for gathering of information on your basic clothing items. Then take the time to treat yourself to a leisurely lunch. Find a place with nice surroundings where service is personal. Relax and look at your list. There is no pressure on you to buy anything, just to contemplate. If you find an item which suits your wardrobe plan, and the colour, the style and the price are right, then buy it that day. Otherwise, take another two hours of looking and enjoying yourself.

Before you go home, purchase something special just for you: perfume, a nice cake of soap, bath oil, a pretty handkerchief or a bunch of flowers. Wouldn't it be nice if you could arrange to have dinner out? Or maybe having dinner ready at home will help to complete this rewarding day.

The Shopping Day

When you have digested the information gathered on your browsing day, and you now have a precise shopping list, your shopping day will be one of accomplishment. Dress in your most business-like costume (City) which will communicate self-assurance and confidence. You have made the decision to purchase those well-chosen items for your wardrobe. You need no more assistance. You are ready to purchase. Clothes are only one of the good things in life to enjoy and provide some of the greatest pleasures—like good friend, good music, good wine, good food and wonderful colours. Feel good about yourself and the prospect of having a colour-coordinated wardrobe. Before you venture home, pick up some special little gifts—this time for the family. A magazine for him, colouring pencils for the kids and a special dessert for all. You will be so well organized that you could be ready for a trip around the world—all it takes is to pack one suitcase.

Some Final Reminders

Review your Colour, Texture and Style preferences for your clothing
Have you decided on a colour scheme you intend to use which suits your personality? Will you use Colour, Texture and Style to convey a specific message? Make sure you are not going all over the spectrum of colours. Remember, you want a colour-coordinated wardrobe to go an a trip around the world and be able to pack it all in one suitcase.

Review the accessories in your home
Does your home look anything like you? Does it have your signature? Is there room in your decor to convey that another person lives there with you? Buy some pillows, place mats and candles. Have a date!

Check the garments for your stage
Is your wardrobe suited to your lifestyle or do you have more clothes taking up space that belong to yesterday? Have fun making a special effort to dress for dinner, for yourself, your partner and your family.

Organize your closet
When you open the door to your closet are you looking at a display of colour-coordinated clothes? Please take the time to organize your closet. It is an uplifting experience to see an uncluttered, well organized closet.

The Little Things in Life

I DO NOT BELIEVE THAT RESPECT AND COURTESY SHOULD BE FASHION items, doomed to go in and out of style. Do You? We live in a time where everybody is involved in many activities. To take advantage of all opportunities to show thoughtfulness, based on genuine kindness and consideration, creates a bond between you and the people with whom you come in contact.

Truly, it is the little things in life that provide opportunities to enhance your quality of life.

Be Kind to each other
Give hugs
Bring flowers
Open doors for him, for her, for everyone
Say thank you a lot, a thousand times
Write thank you notes
Say thank you
Write love notes to your husband and mail them
Write love notes to your children and mail them
Send flowers to your husband's, wife's or partner's office
Have a special date with your husband, your partner or friend every month
Say thank you
Dress with respect for others
Smile at sales people
Sales people—smile at customers
Say thank you for a job - any job - well done

Give compliments to family, friends and strangers
Dress with consideration for the occasion and the venue
Give praise to your partner and friends
Bring flowers
Write thank-you notes for a special moment
Acknowledge a pretty thank-you note at the next opportunity
Have candlelight dinners with your children
Buy yourself fresh flowers often
Have a picnic by the fireplace
Bring flowers to you neighbour, "just because"
Be an example of courtesy to others
Say thank you
Accept a compliment with a "thank you"
Dress up for Sunday family dinner
Give a phone call to your out of town friend once a month
Say thank you from the bottom of your heart
And Thank God Every Day!

My Bunch of Thank You's

With sincere gratitude I thank my heavenly partner for giving me a creative mind, the dream to write this book, the inspirations and for answering my prayers for a wonderful support group. From the bottom of my heart I thank my dear partner here on earth, my husband Francis, for being with me every step of the way, as a reader, editor, consultant and supporter on the exciting journey of writing this book.

Many thanks to my daughter Fiona who dreamed the dream with me, for her intellectual support; and for the typing of my original notes over the years.

I thank Lynn Jambor who so kindly and in good spirit assembled all my manuscript and transfered it onto the computer.

My sincere thank you to Evan Warneboldt who so patiently produced all the beautiful pictures. I thank my book designer Catherine Jordan who created this book from my text and photographs.

A warm embrace to the wonderful people who so willingly opened their homes for the photographs: Joan and Bill Ritchie, Karen Matheson, Deborah Serin and June Matheson.

I also thank Maureen Sullivan of Thomas Hobbs for the beautiful country bouquet on the cover of the book and Shelley Owen of Desdemona's for the lovely flowers for the Black personality. A warm thank you to my friends Joan Ritchie who created the flower arrangements for the other personalities and Doris McFadyen for draping the dressform in silk and velvet and to Eleanor Jones for proofreading. My sincere thank you to Holt Renfrew for providing the clothing, to Louis Vuitton for the luggage, to Chintz & Company for the dress form and to The Country Store for accessories.

Last but by no means least, I thank my friend Samaya Ryon who by giving me the *Artist's Way* by Julia Cameron made all the above activities and accomplishments happen. Thank you for your loving support.

I thank you all! Rohita